ANNE C. GUTTERIDGE

BARNES & NOBLE THESAURUS OF BIOLOGY

the principles of biology
explained and illustrated

BARNES & NOBLE BOOKS
A DIVISION OF HARPER & ROW, PUBLISHERS
New York, Cambridge, Philadelphia, San Francisco
London, Mexico City, São Paulo, Sydney

Library of Congress Cataloging in Publication Data

Gutteridge, Anne.
    Barnes & Noble thesaurus of biology.

    Includes index.
    1. Biology—Dictionaries. I. Title. II. Title: Thesaurus of biology.
QH302.5.G87   1983   574'.03'21   83-47594
ISBN 0-06-015213-3
ISBN 0-06-463581-3 (pbk.)

This book was designed and produced by
BLA Publishing Limited, The Studio,
Newchapel Road, Lingfield, Surrey, England.

Illustrations by Rosie Vane-Wright
Phototypeset in Britain by Composing Operations Limited
Colour origination by Chris Willcock Reproductions
Printed in Belgium

Contents

How to use this book

This book contains some 2700 words used in biology. These are arranged in groups under the main headings listed on pp.3–4. The entries are grouped according to the meaning of the words to help the reader to obtain a broad understanding of the subject.

At the top of each page the subject is shown in bold type and the part of the subject in lighter type. For example, on pp.12 and 13:

12 · CELLS/OSMOSIS

CELLS/STRUCTURE · **13**

In the definitions the words used have been limited so far as possible to words in common use.

---

1. To find the meaning of a word

Look for the word in the alphabetical index at the end of the book, then turn to the page number listed.

The description of the word may contain some words with arrows in brackets (parentheses) after them. This shows that the words with arrows are defined near by.

(↑) means that the related word appears above or on the facing page;

(↓) means that the related word appears below or on the facing page.

A word with a page number in brackets (parentheses) after it is defined elsewhere in the dictionary on the page indicated. Looking up the words referred to in either of these two ways may help in understanding the meaning of the word that is being defined. For example, the entry *erepsin* on p.112 is

erepsin (*n*) a mixture of proteolytic enzymes (p.111); mainly splits peptides into amino acids. It is found in the succus entericus (↑) of man.

In order to understand what *erepsin* is, reference must be made to *proteolytic enzymes* on p.111 and *succus entericus* which is defined on p.112 above erepsin.

The entry in brackets after a word indicates the part of speech: (*n*) indicates a noun, (*adj*) an adjective, (*v*) a verb.

In the index, occasionally the page reference is in brackets. This indicates the page on which the word is to be found and that it does not have a separate entry. For example: monomer (241). This is found in bold type on p.241 under the heading *polymer*.

2. To find related words

Look in the index for the word you are starting from and turn to the page number shown. Because this book is arranged by ideas, related words will be found in a set on that page or one near by. The illustrations will also help here.

For example an *enzyme* is defined on p.109. Words relating to enzymes, groups of enzymes and examples of enzymes are found on this and following pages (p.110-112).

3. As an aid to studying or reviewing

There are two methods of using this book in studying or reviewing a topic. You may wish to see if you know the words used in that topic or you may wish to review your knowledge of a topic.

(*a*) To find the words used in connection with *lipids* look up *lipid* in the alphabetical index. Turning to the page indicated, p.123, you will find *lipid, fat, wax, steroid*. Turning over to p.124 you will find more related words.

(*b*) Suppose you wished to review your knowledge of a topic, e.g. *manufacture of proteins*. If, say, the only term you could remember was *protein* you could look it up in the alphabetical index. The page reference is to p.119. There you would find the words *protein, protein synthesis, peptide, polypeptide,* etc. If you next look at the words you are referred to under these entries this will lead to a fuller understanding of the topic.

4. To find a word to fit a required meaning

It is almost impossible to find a word to fit a meaning in most dictionaries, but it is easy with this book. For example, if you had forgotten the word for the substances which make up RNA, all you would have to do would be to look up RNA in the index and turn to the page indicated, p.118. There you would find the word you wanted which is *nucleotides* and also related words such as *nucleoprotein, nucleic acid*.

biology (n) study of living things.

taxonomy (n) study of the classification (p.18) of living things, mostly according to their structural similarities and differences.

botany (n) study of plants.

zoology (n) study of animals.

anthropology (n) study of man.

cytology (n) study of cells using light, and electron, microscopes. *Compare histology* (↓), *morphology* (↓), *anatomy* (↓).

histology (n) study of the structure of tissues at the cellular level made by microscopic examination of stained sections. *Compare cytology* (↑), *morphology* (↓), *anatomy* (↓).

morphology (n) study of the structure of living things, mainly at the cellular and tissue level. *Compare cytology* (↑), *histology* (↑), *anatomy* (↓).

anatomy (n) study of the macroscopic (p.10) structure of living things, mainly at the organ and tissue level. *Compare cytology* (↑), *histology* (↑), *morphology* (↑).

physiology (n) study of how organisms (p.84) and parts of organisms function and the processes involved.

neurophysiology (n) physiology (↑) of the nervous system.

histochemistry (n) study of the chemical aspects of histology (↑).

biochemistry (n) study of large and small molecules (p.238) and the chemical reactions that occur in living things.

molecular biology (n) study of the structure and functioning of large molecules (p.238) involved in living processes.

endocrinology (n) study of the structure and functioning of endocrine glands (p.180) and the hormones (p.180) they secrete.

immunology (n) study of antigens (pp.141, 222), antibodies (pp.142, 223) and their interactions.

embryology (n) study of the development of embryos (pp.76, 194).

pathology (n) study of diseased and other abnormal tissue.

genetics (n) study of heredity (p.197) and variations in living things.

microbiology (n) study of microscopic organisms, i.e. Bacteria, Fungi, viruses, parasites, Protozoa.

parasitology (n) study of parasites, especially Protozoa (p.21) and worms (p.31).

protozoology (n) study of Protozoa (p.21).

bacteriology (n) study of Bacteria (p.27).

entomology (n) branch of zoology; the study of insects (p.37).

ecology (n) study of the relationship of plants and animals to their environment (p.227).

palaentology (n) study, through fossils, of how living things relate to their environment in the past.

geology (n) study of the earth's crust.

microscope (*n*) an instrument used to magnify (↓) objects too small to be seen with the naked eye. The ordinary laboratory microscope, the compound microscope has two sets of lenses, objective and eye piece. Maximum magnification is approximately × 1500.

phase contrast microscope (*n*) a microscope (↑) which obtains clear detail of transparent specimens without the need for staining (↓) using differences in the phase of the light rather than wavelength.

ultra-violet microscope (*n*) a microscope (↑) which uses ultra-violet radiation as the light source. It has greater resolution (↓) than the light microscope.

electron microscope (*n*) a microscope (↑) which passes electrons instead of light through the sample. It has a higher resolution (↓) than the light microscope. Magnifications of × 250000 are obtained.

micrograph (*n*) the photograph of an image produced by a light or electron microscope (↑).

resolution (*n*) the smallest distance between two objects before they can be distinguished as separate objects. Two objects which cannot be seen separately are beyond the **limit of resolution** of a microscope (↑).

magnify (*v*) to make greater; to increase in size.

magnification (*n*) the amount of size increase.

micrometre (*n*) unit used for measuring microscopic length. It equals one-thousandth of a millimetre or 10^{-6} metre. Symbol: μm.

nanometre (*n*) unit of microscopic length equal to one-thousandth of a micrometre (↑) or 10^{-9} metre. Symbol: nm.

micron (*n*) former unit of microscopic length, it is equal to the micrometre (↑). Symbol: μ.

Ångström (*n*) unit formerly used to measure microscopic length. One Ångström equals 10^{-10} metre; ten Ångström units equal one nanometre (↑). Symbol: Å.

ultrastructure (*n*) the detailed structure of a cell which can only be seen using an electron microscope (↑).

slide (*n*) a thin, oblong piece of glass on which sections are placed for microscopic (↑) examination.

cover slip (*n*) a very thin piece of glass, about 0.2mm thick, used to cover sections (↓) which have been prepared, e.g. in Canada balsam (p.10), for microscopic examination.

specimen (*n*) an object or part of an object which is used as a sample for examination or study.

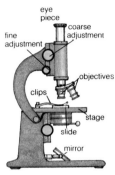

compound microscope

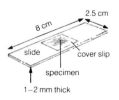

slide

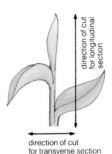

direction of cut for longitudinal section

direction of cut for transverse section

stem sections

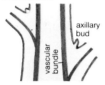

axillary bud

vascular bundle

longitudinal section of a stem

vascular bundle

transverse section of a stem

axis

radial longitudinal section

longitudinal section

axis

sagittal section

axis

tangential longitudinal section

section (*n*) a very thin slice of tissue prepared for examination under a microscope. Sections may be fixed (↓) and stained (↓).

longitudinal section, L.S. (*n*) a section which is cut along the length of a specimen (↑). If along line of radius it is a **radial longitudinal section**, if parallel to a line passing through the axis it is a **sagittal section**; if neither it is a **tangential longitudinal section**.

transverse section, T.S. (*n*) a section which is cut across a specimen.

fixation (*n*) first step in the preparation of a sample for microscopic examination. Fixation kills the cells but preserves their original shape. The most common **fixatives (fixation agents)** are alcohol, gluteraldehyde and osmium tetroxide. After fixing, the sample is dehydrated (p.10), cleared (p.10) and cut into sections using a microtome (p.10).

staining (*n*) the colouring of specimens on microscope slides for examination. Some stains colour all the specimen, others are only taken up by certain tissues. Common stains include eosin Y, gentian violet, Giemsa, aniline blue.

stain	examples of tissues stained
eosin Y	cellulose, cytoplasm
gentian violet	nuclei
Giemsa	blood, blood parasites
aniline blue	Fungi, sieve plates
methylene blue	bacteria, nuclei
Leishman's stain	blood cells, nuclei of white blood cells

vital staining (*n*) the colouring of cells, while they are still living, by harmless dyes. Stains used include methylene blue, Janus green.

Leishman's stain (*n*) a stain (↑) used to colour blood cells for microscopic examination.

Feulgen stain (*n*) a stain (↑) which colours DNA purple.

dehydration (*n*) a process of gradual removal of water from a tissue for microscopic examination by soaking it in increasing concentrations of alcohol. It is essential because embedding wax and Canada balsam (↓) do not mix with water.

clearing (*n*) process of soaking a tissue in xylene or benzene after dehydration (↑) in order to make it transparent. Used in preparation of slides (p.8).

microtome (*n*) an instrument for cutting very thin slices (a few micrometres (p.8)) of specimens for microscopic examination. Sections are cut from the frozen sample or by embedding the sample in wax for support.

Canada balsam (*n*) a gum from the balsam fir tree which is dissolved in xylene and used to make permanent preparations of samples on microscope slides. The section (p.9) is placed between the slide (p.8) and the cover slip in a thin layer of Canada balsam which dries hard, with the same refractive index as glass.

immersion-oil (*n*) oil of the same refractive index as glass; used to improve resolution (p.8). It fills the space between the cover slip and objective lens of a microscope.

microscopic (*adj*) describes objects which can only be seen with the aid of a microscope (p.8). *Compare macroscopic* (↓).

macroscopic (*adj*) describes objects which can be seen with the naked eye, e.g. a petal, an earthworm. *Compare microscopic* (↑).

multicellular (*adj*) describes organisms or parts of organisms made up of many cells. *Compare unicellular* (↓).

unicellular (*adj*) describes organisms made up of one cell, e.g. *Amoeba*, a protozoon (p.21). *Compare multicellular* (↑).

multinucleate, polynucleate (*adj*) containing more than one nucleus, e.g. striated muscle.

mononucleate (*adj*) describes a cell which contains only one nucleus, e.g. a lymphocyte (p.143, 223).

extracellular (*adj*) describes materials or processes in an organism that occur outside the plasma membrane (p.13) of the cell. *Compare intracellular* (↓), *intercellular* (↓).

intracellular (*adj*) describes materials or processes in an organism that are within the plasma membrane (p.13) of the cell. *Compare extracellular* (↑), *intercellular* (↓).

unicellular

Amoeba, a unicellular organism

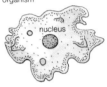

multinucleate

muscle fibre, multinucleate

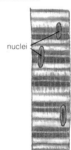

mononucleate

lymphocyte, a mononucleate cell

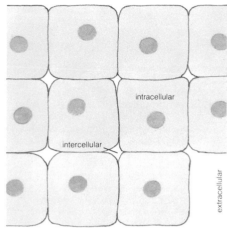

intercellular (*adj*) describes materials or processes that occur in the space between cells of tissues or organs of multicellular organisms, e.g. intercellular air-spaces in plants are cavities between adjacent cells.

subcellular organisation (*n*) the division of the cell into organelles (p.13), e.g. ribosomes, mitochondria.

intercellular fluid, interstitial fluid, tissue fluid (*n*) fluid which fills spaces between the cells of the tissues in animals. Its composition is kept constant (homeostasis, p.153); differences cause cell damage.

internal environment (*n*) environment surrounding tissue cells.

differentiation (*n*) change of unspecialised cells into those that have a particular structure and/or function, e.g. embryo cells become differentiated into specialised cells of the various tissues and organs.

hypertrophy (*n*) increase in size of a tissue or organ by enlargement of its cells. *Compare hyperplasia* (↓).

hyperplasia (*n*) increase in size of a tissue or organ due to cell division. *Compare hypertrophy* (↑).

atrophy (*n*) decrease in amount of tissue or size of an organ; often caused by lack of use.

autolysis (*n*) digestion of cells and tissues by their own enzymes.

cytolysis (*n*) bursting of the cell membrane. It causes the death of the cell.

osmosis (*n*) passage of solvent molecules across a semi-permeable membrane (↓) from a less concentrated to a more concentrated solution.

osmotic pressure (*n*) pressure that has to be applied across a semi-permeable membrane (↓) to stop osmosis (↑) occurring.

osmotic potential (*n*) pressure which if applied will prevent osmosis (↑) by any particular solution.

diffusion pressure deficit (*n*) force with which water enters a cell.

semi-permeable membrane (*n*) membrane which allows passage of solvent but not solute molecules. In osmosis (↑) it allows the passage of solvent, but not sugars and salts.

impermeable membrane (*n*) membrane which will not allow passage of solute or solvent molecules.

permeable membrane (*n*) membrane which allows passage of solute and solvent molecules from regions of high to low concentrations.

permeability (*n*) degree to which a membrane allows a particular molecule to pass across it.

hypertonic (*adj*) a solution is said to be hypertonic relative to another, if, when the two are separated by a semi-permeable membrane (↑), water flows into it; if water flows out, it is **hypotonic** (*adj*) to the other solution. If no water flows, solutions are said to be **isotonic** (*adj*). The movement of water is by osmosis (↑).

plasmolysis (*n*) shrinkage of protoplasm from a plant cell wall when the cell is in a hypertonic (↑) environment. Due to water loss by osmosis (↑); turgor (↓) is decreased.

turgor (*n*) state of a cell when osmosis (↑) has caused the cell volume to increase to its maximum, making the cells turgid. Turgor maintains the shape of the cell and provides mechanical support for the plant.

physiological saline (*n*) solution of sodium chloride in water (0.9% by weight). It is isotonic (↑) with cells and can keep them alive for short periods. *Compare Ringer's solution* (↓).

Ringer's solution (*n*) aqueous solution of mainly sodium, potassium and calcium chlorides used to keep cells and tissues alive during *in vitro* experiments. The salt concentrations used are identical to those in intercellular fluid (p.11). Ringer's solution when buffered to the correct pH (p.240) keeps cells alive for a long time. *Compare physiological saline* (↑).

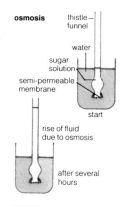

osmosis

thistle— funnel

water

sugar solution

semi-permeable membrane

start

rise of fluid due to osmosis

after several hours

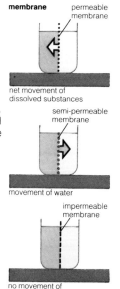

membrane

permeable membrane

net movement of dissolved substances

semi-permeable membrane

movement of water

impermeable membrane

no movement of solute or solvent

☐ lower concentration

☐ higher concentration

cell (*n*) basic unit of all living organisms except the viruses. Cells differ in size, shape and function. All contain protoplasm (↓) and are surrounded by a membrane (↓); most contain a nucleus. New cells are formed from existing cells by division or fusion.

cellular (*adj*) made up of, or containing, cells.

prokaryote (*n*) a unicellular organism (e.g. a bacterium) which lacks a true nucleus and other membrane-bound organelles (↓). Its DNA (p.208) lies free in the cytoplasm (↓). *Contrast eukaryote* (↓).

eukaryote (*n*) a unicellular or multicellular organism whose cells contain DNA (p.208) in a nucleus enclosed by a membrane (↓) and which contains in its cytoplasm (↓) membrane-bound organelles (↓) such as mitochondria. *Contrast prokaryote* (↑).

membrane (*n*) a thin layer (5–10 nanometres thick) surrounding organelles (↓) and cells (↑); it controls movement of substances in and out. It consists mainly of phospholipids and proteins, the proportions of which vary with position and function of the membrane.

plasma membrane, cell membrane (*n*) membrane surrounding a cell; damage to it destroys the cell.

cell wall (*n*) the layer of prokaryote (↑) and plant cells external to the plasma membrane (↑). In plants it consists mainly of cellulose which provides mechanical support. Animal cells do not have a cell wall.

protoplasm (*n*) the living contents of a cell, i.e. the plasma membrane (↑), cytoplasm (↓) and nucleus. *Compare cytoplasm* (↓).

cytoplasm (*n*) the living contents of a cell excluding the plasma membrane (↑), vacuoles and nucleus. It is a viscous fluid surrounded by the plasma membrane and contains organelles such as mitochondria (p.14), Golgi apparatus (p.14), ribosomes. *Compare protoplasm* (↑).

cytosol (*n*) the soluble part of the cytoplasm (↑) which remains after removal of the organelles (↓).

ectoplasm (*n*) the thin, outer, gel-like layer of cytoplasm (↑), especially of animal cells. Unlike endoplasm (↓) it lacks organelles (↓).

endoplasm (*n*) the inner layer of cytoplasm (↑), especially of animal cells. It is less solid than ectoplasm (↑) and contains many organelles (↓).

organelle (*n*) specialised (p.15) part of a living cell. It is often surrounded by a membrane, e.g. nucleus, chloroplast, mitochondrion.

plant cell

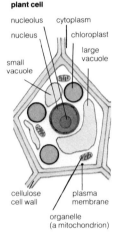

nucleolus   cytoplasm
nucleus   chloroplast
small   large
vacuole   vacuole
cellulose   plasma
cell wall   membrane
organelle
(a mitochondrion)

nucleus (*n*) a membrane-bound organelle (p.13) in a
eukaryote (p.13) cell which contains the DNA complex-
ed with protein to form the chromosomes. It is essential
for the long term survival of the cell. Nuclei are normally
formed from existing nuclei by mitosis or meiosis (p.201).

nucleolus (*n*) a small body present in resting nuclei.
RNA is made in it.

centriole (*n.pl.*) two small granules found just outside the
nucleus (↑) in most animal and lower plant cells. They
replicate before mitosis. The four centrioles thus formed
move in pairs to opposite poles of the mitotic spindle.

centrosome (*n*) the area of the cytoplasm that contains
the centriole (↑); it stains darkly.

vacuole (*n*) a fluid-filled space inside a cell surrounded
by a membrane (p.13). Plant cells often have one large
vacuole surrounded by a membrane called a **tono-
plast**; it contains cell sap (↓) and controls turgor (p.12).
In animal cells there are usually many, smaller vacuoles.

cell sap (*n*) a solution of organic and inorganic
substances found in the vacuoles (↑) of plant cells. It is
isotonic (p.12) with protoplasm (p.13).

endoplasmic reticulum, E.R. (*n*) a membrane system
found in the cytoplasm of most eukaryotic (p.13) cells. If
ribosomes (↓) are attached to its surface, it is referred to
as rough endoplasmic reticulum; if not as smooth.

ribosome (*n*) a small particle found in the cytoplasm
(p.13) of living cells. It consists of ribosomal RNA
(rRNA) and protein. It is involved in protein synthesis.

polysome, polyribosome (*n*) the structure formed when
several ribosomes (↑) are associated with a molecule of
messenger RNA (mRNA) during protein synthesis.

mitochondrion (*n*) (*pl.mitochondria*) organelles found in
cell cytoplasm (p.13) (except bacteria and blue-green
algae); contain enzymes and carry out aerobic respiration.

Golgi apparatus, Golgi body (*n*) organelle found in the
cytoplasm (p.13) of most animal and some plant cells. It
is concerned with export of proteins from the cell.

lysosome (*n*) organelle found in the cytoplasm of
eukaryote (p.13) cells; contains enzymes which can
cause autolysis (p.11).

plastid (*n*) organelle found in plant cytoplasm. Plastids
are classified according to colour. **Leucoplasts**
(involved in food storage) are colourless.
Chromoplasts contain pigments (p.58), especially
chlorophyll (involved in photosynthesis, p.58).

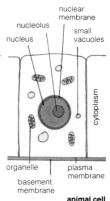

animal cell

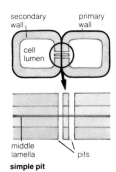

secondary wall primary wall

cell lumen

middle lamella   pits

simple pit

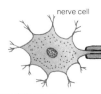

nerve cell

specialised cell

tissue

muscle cells together form muscle tissue

middle lamella (*n*) the intercellular (p.11) layer between adjacent plant cells.

pits (*n.pl.*) small areas in the wall of a plant cell that remain thin to allow movement of substances from cell to cell. They are normally found in pairs on either side of the middle lamella (↑) between neighbouring cells.

specialisation (*n*) the evolution (p.225) of special features in an organism so that it becomes better adapted to its environment (p.227). This is a disadvantage if the environment changes; over-specialisation can then lead to the extinction of the species.

specialised cell (*n*) a cell in a higher animal or plant that has developed special characteristics to enable it to perform particular functions, e.g. bone, muscle and nerve cells.

fibre (*n*) (1) a long sclerenchyma (p.49) cell which provides mechanical support in plants. (2) a fine, thread-like structure found in animal tissues, e.g. nerve fibre, muscle fibre.

tissue (*n*) a large group of cells, mainly of the same kind, which is specialised (↑) to carry out a particular function, e.g. muscle cells form muscle tissue. Tissues joined together form the organs of multicellular organisms.

tissue culture (*n*) a method for keeping small quantities of plant or animal cells and tissues alive after removal from the organism. Cells are suspended in a sterile (p.28), isotonic (p.12) medium at the correct pH and temperature and are supplied with oxygen and nutrients. The medium must be renewed periodically to replenish the food supply and remove waste products. A similar method can be used for keeping organs alive.

connective tissue (*n*) a supporting and packing tissue found in vertebrates. Structures like nerves and blood vessels are embedded in it. It consists mainly of collagen (white fibrous tissue) with some elastin (yellow fibrous tissue), reticular fibres and cells.

mast cell (*n*) a cell found in connective tissue (↑); it has granules which stain strongly and from which histamine (p.217) can be released.

stroma (*n*) intercellular (p.11) material or the connective tissue (↑) part of an animal organ.

epithelium (*n*) tissue (p.15) forming a continuous layer of cells covering both internal (e.g. respiratory tract) and external (e.g. skin) surfaces. Cells are close together with little intercellular (p.11) substance. Epithelia are classified according to shape, e.g. **columnar epithelia** (column-shaped) and **squamous epithelia** (flat cells).

endothelium (*n*) a single layer of cells lining the heart, blood vessels and lymphatic vessels of vertebrates.

basement membrane (*n*) thin intercellular (p.11) membrane lying under most animal epithelial cells. It provides support for the epithelium (↑) and controls exchange of substances from the epithelium to tissues lying below.

ciliated epithelium (*n*) surface epithelium (↑) with cilia (↓) on the exposed surface of the cells. Found, for example, lining respiratory tract and fallopian tubes.

cilium (*n*) (*pl.cilia*) a fine thread-like organelle which projects from the surface of some eukaryote (p.13) cells. Cilia move and beat rhythmically together to transport fluid and particles along the surface of the cells or to move the cell (e.g. *Paramecium*). Cilia are most common in animal cells. They are microscopically similar to flagella (↓) but are shorter and more numerous.

flagellum (*n*) (*pl.flagella*) a long, thread-like organelle projecting from the surface of a cell, e.g. in motile gametes, algae (*Euglena*) and protozoa (*Trypanosoma*). Usually only one and never more than a few are present per cell. It causes movement of the cell. *Compare cilium* (↑).

serous membrane (*n*) a layer of tissue lining many internal vertebrate cavities that have no opening to the outside of the body, e.g. pleural, pericardial and peritoneal cavities. *Compare mucous membrane* (↓).

mucous membrane, mucosa (*n*) the moist epithelium (↑) and its underlying connective tissue (p.15) lining many internal vertebrate cavities that open to the outside of the body, e.g. intestinal, respiratory and genito-urinary tracts. *Compare serous membrane* (↑).

mucus (*n*) (1) a slimy, viscous fluid produced by the epithelial cells of a mucous membrane (↑). (2) any viscous or slimy fluid secreted by invertebrates.

secrete (*v*) act of discharging material made by a cell. Most cells have secretory activity. Gland cells are specialised secretory cells, e.g. sebaceous gland cells secrete sebum.

secretion (*n*) substance secreted by a gland cell, e.g. sebum, hormones, enzymes.

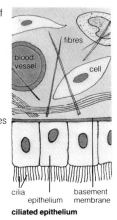

cilia   basement
epithelium   membrane
ciliated epithelium

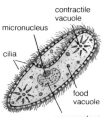

contractile
vacuole
micronucleus

cilia

food
vacuole

meganucleus
cilium
Paramecium

flagellum
Euglena

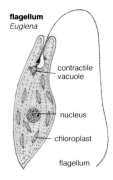

contractile
vacuole

nucleus

chloroplast

flagellum

matrix (*n*) the intercellular (p.11) substance of animals in which cells are embedded, e.g. bone cells are embedded in bone matrix.

cuticle (*n*) non-cellular layer covering an animal or plant. It is secreted by and external to the epidermis. In plants it is made of cutin (↓) and although it protects against mechanical injury its main function is to prevent excess water loss during transpiration (p.47). In invertebrates the cuticle can protect against water loss but it is strengthened and functions mainly as an exoskeleton protecting against mechanical damage.

cutin (*n*) complex substance of which plant cuticle (↑) is made. Impermeable to water, it restricts water loss.

pellicle (*n*) a layer surrounding many unicellular organisms. It is protective, flexible and gives shape.

capsule (*n*) (1) a structure, usually composed of connective tissue, (p.15) surrounding an animal cell, tissue or organ. It provides protection and support. (2) gelatinous material secreted by some bacteria. It forms a protective envelope around the cell.

septum (*n*) dividing wall or partition in an animal or plant structure.

lacuna (*n*) (1) in plants, a cavity inside a tissue or organ that contains water or secreted gases. (2) in animals, any small cavity. *Compare lumen* (↓).

lumen (*n*) the space within a cell, tube or duct. It may be formed by many cells, e.g. lumen of the stomach, or by cells stacked on top of one another, e.g. the cavity of a tracheid or xylem vessel in which water transport occurs. It is also the cavity inside a plant cell which has lost its living contents. *Compare lacuna* (↑).

contractile vacuole (*n*) a cavity in free-living fresh water protozoa that slowly enlarges and then suddenly contracts expelling its contents to the exterior. Its function is to eject excess water drawn into the cell by osmosis.

pseudopodium (*n*) temporary projection of the cell in some Protozoa (p.21) and macrophages (white blood cells). In *Amoeba* it is used for locomotion and feeding.

protrude (*v*) to thrust or push out. **protrusion** (*n*).

project (*v*) to stick out from. **projection** (*n*).

ergastic substances (*n.pl.*) former term for waste products of metabolism stored in protoplasm, vacuoles, cell walls.

interior (*n*) the inside or inner part.

exterior (*n*) the outside or outer part.

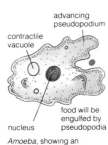

advancing pseudopodium
contractile vacuole
food will be engulfed by pseudopodia
nucleus

Amoeba, showing an example of pseudopodia

classification (*n*) a scientific arrangement of living organisms into groups of similar individuals.

natural classification (*n*) the classification (↑) of living organisms into groups according to similarities and differences (mainly structural) between them. The smallest unit is the species (↓). Species which are alike are grouped into genera (↓), similar genera are grouped into families (↓), then orders (↓), classes (↓) and finally phyla (↓) (animals) and divisions (↓) (plants). Progressing from species to phyla and divisions, the number of similarities between members of a group becomes fewer. *See taxonomy* (p.7), *binomial nomenclature* (↓).

binomial nomenclature (*n*) nomenclature using two names. The international scientific system used today for naming animal and plant species. It was first used by Linnaeus in the eighteenth century. Names, in Latin, are written in italics. The first generic (↓) name has a capital letter; the second, which is specific to that organism, is written with a small letter, e.g. *Homo sapiens*. *See natural classification* (↑), *taxonomy* (p.7).

holotype, type specimen (*n*) the specimen of an animal or plant used to describe a new species (↓); it is preserved for future reference.

kingdom (*n*) a major division of living things, e.g. animal and plant kingdoms.

phylum (*n*) (*pl.phyla*) one of the major units used in the classification (↑) of the animal kingdom (↑), e.g. Arthropoda, Nematoda. It is further divided into classes (↓). *Compare division* (↓).

division (*n*) one of the main units used in the classification (↑) of the plant kingdom (↑), e.g. Thallophyta, Bryophyta. Divisions are further divided into classes (↓). *Compare phylum* (↑).

class (*n*) unit of classification (↑) into which phyla (↑) and divisions (↑) are divided. Classes are further divided into orders (↓).

order (*n*) unit of classification (↑) into which classes (↑) are divided. Orders are further divided into families (↓).

family (*n*) unit of classification (↑) into which orders (↑) are divided. Families are further divided into genera (↓).

genus (*n*) (*pl.genera*) unit of classification (↑) into which families (↑) are divided. Genera are further divided into species (↓).

generic (*adj*) of, or belonging to, a genus (↑).

classification

A. modern man

Kingdom	**Animal**
Phylum	**Chordata**
Sub-Phylum	**Vertebrata**
Class	**Mammalia**
Order	**Primates**
Family	**Hominidae**
Genus	*Homo*
Species	*sapiens*

B. wild rose

Kingdom	**Plant**
Division	**Spermatophyta**
Sub-Division	**Angiospermae**
Class	**Dicotyledonae**
Order	**Rosales**
Family	**Rosaceae**
Genus	*Rosa*
Species	*canina*

species (*n*) unit of classification (↑) into which genera (↑) are divided. It is the smallest unit normally used. It usually refers to a group of individuals unable to breed with other groups of organisms but capable of breeding amongst themselves to produce fertile offspring.

sub-species (*n*) unit of classification (↑) into which species (↑) are further divided. They are formed by the effect of natural selection on geographically isolated populations of the species. If isolation continues, sub-species eventually become species.

specific (*adj*) characteristic or typical of a species (↑).

variety (*n*) unit of classification into which species (↑) are divided. It is mostly used for cultivated forms of a plant.

polymorphism (*n*) the existence of more than one form of individual in the same species (↑).

Thallophyta (*n.pl.*) division of the plant kingdom (↑) which contains the most primitive forms of plant life, the *Algae* (↓). The plant body or **thallus** is unicellular, e.g. *Euglena* (p.16) or multicellular, e.g. sea-weed. There is no differentiation into roots, stems or leaves. Asexual reproduction is by spores; sexual reproduction by fusion of gametes. Fungi (↓) and **slime moulds** are sometimes included in the Thallophyta.

Algae (*n*) sub-division of the Thallophyta (↑). They are usually aquatic and possess chlorophyll, e.g. *Spirogyra*, desmids, sea-weeds, diatoms. Algae are divided into several classes according to their pigments.

Fungi (*n.pl.*) (*sing.fungus*) considered either as a sub-division of the Thallophyta (↑) or as a separate kingdom (↑). All lack chlorophyll and are parasitic or saprophytic, reproducing by spores. Fungi are divided into classes depending on the type of spore they produce. Fungi include rusts, yeasts, bread mould, mushrooms, *Mucor*, *Penicillium*.

Lichen (*n*) a composite organism of an Algae (↑) and a Fungus (↑) living together symbiotically (p.30).

Bryophyta (*n.pl.*) division of the plant kingdom (↑). It is a group of small, multicellular green plants with widely differing habitats. The plant body is differentiated into stems and leaves; lacks vascular tissue; has root-like rhizoids. Bryophyta show alternation of generations; the main plant form is the gametophyte, the sporophyte is dependent on it. They are divided into Hepaticae or liverworts and Musci or mosses.

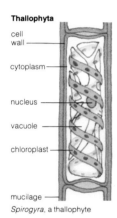

Thallophyta

cell wall

cytoplasm

nucleus

vacuole

chloroplast

mucilage

Spirogyra, a thallophyte

Bryophyta

capsule or spore case

sporophyte generation

leaf

gametophyte generation

rhizoid

moss, a bryophyte

Pteridophyta (*n.pl.*) division of the plant kingdom (p.18). They are flower-less, vascular plants with roots, stems and leaves. Pteridophyta are divided into several classes, e.g. **ferns**. They show alternation of generations; the leafy fern is the sporophyte, the gametophyte is small and insignificant but free. Many are extinct.

Spermatophyta (*n.pl.*) division of the plant kingdom (p.18). They are seed-bearing plants with stems, leaves, roots and flowers which form fruits and seeds. They show alternation of generations; the main plant is the sporophyte, the gametophyte is much reduced and dependent on it. Spermatophyta are divided into Gymnospermae (↓) and Angiospermae (↓).

Gymnospermae (*n.pl.*) sub-division of the Spermatophyta (↑). They are plants with naked ovules and seeds (i.e. not inside an ovary), often found on scales of cones. They do not possess true flowers, e.g. the pine, a conifer.

Pteridophyta

leaf or frond

part of leaf or frond, underside showing spores

a pteridophyte, fern

roots

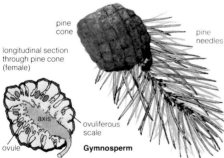

pine cone

pine needles

longitudinal section through pine cone (female)

axis

ovuliferous scale

ovule

Gymnosperm

Angiospermae, flowering plants (*n.pl.*) a sub-division of the Spermatophyta (↑). Seeds are enclosed in a fruit or nut. Angiosperms are divided into two classes, Monocotyledoneae (↓) and Dicotyledoneae (↓).

Monocotyledoneae (*n.pl.*) class of Angiospermae (↑) which have one cotyledon in the seed. Leaves usually have parallel veins, e.g. grasses, corn. *Compare Dicotyledoneae* (↓).

Dicotyledoneae (*n.pl.*) class of Angiospermae (↑) which have two cotyledons in the seed. Leaves usually are net-veined, e.g. pea, buttercup. *Compare Monocotyledoneae* (↑).

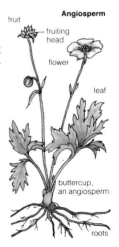

Angiosperm

fruit

fruiting head

flower

leaf

buttercup, an angiosperm

roots

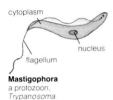

cytoplasm

nucleus

flagellum

Mastigophora
a protozoon.
Trypanosoma

Protozoa (*n.pl.*) considered either as a phylum of the animal kingdom (p.18) or as a separate kingdom. They are simple unicellular organisms living singly or in colonies. They are either free-living in fresh water, salt water or damp terrestrial environments or they live as parasites in other living organisms. Protozoa consist of small units of protoplasm containing at least one nucleus and organelles (e.g. cilia, food vacuoles) specialised to carry out specific functions. Reproduction is normally by binary fission, e.g. *Amoeba, Paramecium, Plasmodium, Entamoeba*. Protozoa are divided into several classes: Sporozoa (↓), Ciliata (↓), Mastigophora (↓), Sarcodina (↓).

Sporozoa (*n.pl.*) class of parasitic Protozoa (↑). They lack organelles of locomotion; food is absorbed from the surroundings over the whole body surface. Their life cycle involves alternation of sexual and asexual reproduction; they produce numerous spores, e.g. *Plasmodium*.

Ciliata (*n.pl.*) class of Protozoa (↑), all of which have cilia (p.16) at some stage of their life cycle. Most have two nuclei (a macronucleus (↓) and a micronucleus (↓)); a mouth; definite organelles, e.g. *Paramecium*.

meganucleus, macronucleus (*n*) a large nucleus found in Ciliata (↑). It divides into equal masses during fission. *Compare micronucleus* (↓).

micronucleus (*n*) a small nucleus found in Ciliata (↑); it divides mitotically. *Compare meganucleus* (↑).

Mastigophora (*n.pl.*) class of Protozoa (↑) that move by one or more flagella (p.16), e.g. *Trypanosoma* (the cause of sleeping sickness).

Sarcodina (*n.pl.*) class of Protozoa (↑) that move and capture food by means of pseudopodia, e.g. *Amoeba* and *Entamoeba* (cause of amoebic dysentery).

Protista (*n*) a grouping of all the unicellular organisms into a separate kingdom distinct from the plant and animal kingdoms. It includes Algae, Fungi, Protozoa and Bacteria which are otherwise considered as part of the plant and animal kingdoms.

Metazoa (*n.pl.*) animals whose bodies are composed of many cells grouped together to form tissues and organs. Not included are Protozoa which are unicellular and sponges which consist of many cells but in a different plan to the Metazoa.

Porifera (*n.pl.*) phylum of the animal kingdom (p.18). They are multicellular animals found mostly in salt water where they are attached to rocks, e.g. sponge.

Ciliata *Paramecium*

micronucleus

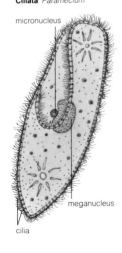

meganucleus

cilia

Coelenterata (*n.pl.*) phylum of the animal kingdom (p.18). They are aquatic, mostly marine animals. The body is made up of two layers, the ectoderm and endoderm surrounding a central cavity that has one opening, the mouth, e.g. *Hydra*, corals, jellyfish, sea-anemonies.

Platyhelminthes, flatworms (*n.pl.*) phylum of the animal kingdom (p.18). The body is composed of three layers of cells, the ectoderm, mesoderm and endoderm. It is unsegmented and flattened providing a large surface area for respiration, e.g. tapeworm *(Taenia)*, liver-fluke, schistosome.

helminth (*n*) a worm which is usually parasitic. *See Platyhelminthes* (↑), *Nematoda* (↓), *Annelida* (↓).

Nematoda, roundworms (*n.pl.*) phylum of the animal kingdom (p.18). They are found in fresh water, salt water, soil and in other living organisms. The body is long, cylindrical, unsegmented and covered with a cuticle. The digestive tube is open at both ends, e.g. filaria, hookworm.

Annelida, segmented worms (*n.pl.*) phylum of the animal kingdom (p.18). The body is divided into many segments each of which has a large coelom, thus they are more complex than nematodes (↑) and platyhelminths (↑), e.g. earthworm.

Mollusca (*n.pl.*) phylum of the animal kingdom (p.18). The body is soft, unsegmented and divided into a head and a muscular foot for movement. The body is covered with a hard shell. They are mostly aquatic, e.g. snail, mussel, oyster.

Arthropoda (*n.pl.*) the largest phylum of the animal kingdom (p.18). The body has jointed, paired legs and a hard exoskeleton of chitin or lime, e.g. crab, lobster, mite, tick, insects like the aphid, spider. The phylum is divided into several classes some of which are Crustacea (↓), Myriapoda (↓), Insecta (↓), Arachnida (↓).

Crustacea (*n.pl.*) class of Arthropoda (↑). They are mostly aquatic. The body is divided into head, thorax, abdomen; has many paired limbs. The head has two pairs of antennae, e.g. lobster, crayfish.

Myriapoda (*n.pl.*) class of terrestrial Arthropoda (↑). The body is long and composed of many segments. Myriapoda have many pairs of walking legs; one pair per segment in **centipedes**, two pairs per segment in **millipedes**.

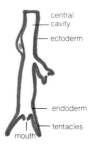

Coelenterata
longitudinal section of a coelenterata, *Hydra*

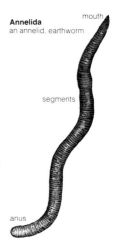

Annelida
an annelid, earthworm

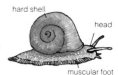

Mollusca
a mollusc, snail

Lepidoptera
a lepidopteran, butterfly

Insecta (*n.pl.*) class of Arthropoda (↑). They are mostly terrestrial. The body is divided into head, thorax and abdomen. The head has one pair of antennae and a pair of compound eyes. The thorax has three pairs of legs. Most have wings; e.g. locust, bug, louse, flea, butterfly, beetle, bee, ant, earwig, termite, fly.

Diptera (*n*) order of Insecta (↑). They have only one pair of wings and a pair of balancers (halteres), e.g. housefly, mosquito.

Lepidoptera (*n.pl.*) order of Insecta (↑), includes butter-flies and moths. They have two pairs of large wings.

Caterpillars are the larvae of butterflies.

caterpillar larva of a butterfly

Arachnida
spider

Arachnida (*n.pl.*) class of Arthropoda (↑). They are mostly terrestrial. The body is divided into two parts; antennae are absent; eyes are simple; there are four pairs of legs, e.g. spider, scorpion, tick, mite.

Echinodermata (*n.pl.*) phylum of the animal kingdom (p.18). They have a marine habitat. Many have external spines. Limestone plates in the skin join to form a skeleton, e.g. starfish, sea-urchin.

Chordata (*n.pl.*) phylum of the animal kingdom (p.18). They have a primative backbone; a hollow nerve cord; gill slits. The phylum is divided into sub-phyla which includes the Vertebrata (Craniata) (↓).

Vertebrata, Craniata (*n.pl.*) sub-phylum of the Chordata (↑). They differ from other Chordata in having a skull which surrounds a well developed brain and a skeleton of cartilage or bone of which a vertebral column forms the central axis. Vertebrata are divided into several classes: Pisces (↓), Amphibia (p.24), Reptilia (p.24), Aves (p.24), Mammalia (p.24).

Invertebrata (*n.pl.*) all animals which are not members of the Vertebrata (↑), e.g. sponge, worm, snail, fly.

Pisces, fish (*n.pl.*) class of aquatic Vertebrata (↑). The body is covered in scales and moves by means of a tail and fins. Fish breathe by means of gills and have a two-chambered heart.

moist
skin

pentadactyl
limb

Amphibia

an amphibian, frog

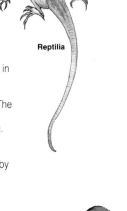

lizard, a reptile

Reptilia

Amphibia (*n.pl.*) class of Vertebrata (p.23). Adults live in damp terrestrial habitats but must lay their eggs in water. Aquatic larvae use gills for breathing; after metamorphosis the adults have lungs for breathing. The body is covered by moist skin; has four pentadactyl limbs; a three-chambered heart; is cold-blooded, e.g. frog, newt.

Reptilia (*n.pl.*) class of Vertebrata (p.23). They mostly live on land; eggs are laid on land and are protected by a shell. The body, which has pentadactyl limbs, is covered with dry, scaly skin to minimise water loss. Lungs are present for breathing throughout life, e.g. snake, alligator, turtle, crocodile.

Aves, birds (*n.pl.*) class of Vertebrata (p.23). The body has pentadactyl limbs and is covered with feathers; bones are light and front limbs modified to form wings for flight. Birds breathe by means of lungs; have a four-chambered heart and are warm-blooded; lay eggs which are protected by a shell.

Aves

wing

a bird, member of
the class Aves

Mammalia (*n.pl.*) class of Vertebrata (p.23). The body is covered with hair or fur; has a diaphragm which separates the chest from the abdominal cavity and is used in respiration. They are warm-blooded; have a four-chambered heart; relatively large brain; young are born alive and fed on milk from mammary glands. The class is divided into many orders including Marsupialia (↓), Insectivora (↓), Cetacea (↓), Carnivora (↓), Rodentia (↓), Primates (↓).

Marsupialia (*n.pl.*) order of Mammalia (↑). Young are born very immature and receive protection and nourishment in a pouch. They are found only in Australasia and in North and South America; e.g. opossum, kangaroo, koala bear.

Insectivora (*n.pl.*) order of Mammalia (↑). They are generally nocturnal and are omnivorous or insect-eating, e.g. mole, shrew, hedgehog.

Cetacea (*n.pl.*) order of Mammalia (↑). They are marine animals in which hind limbs are absent; forelimbs are modified as flippers for stabilization; a blubber layer is found under the skin for insulation; blowholes are present for respiration, e.g. whale, dolphin.

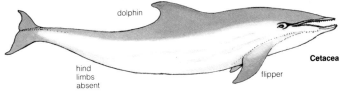

dolphin

Cetacea

hind limbs absent

flipper

Carnivora (*n.pl.*) order of Mammalia (↑). They are flesh eating with strong jaws; teeth are specialised for biting and tearing; claws are well developed to help kill prey. Carnivora are both aquatic, e.g. seal, sea-lion and terrestrial, e.g. lion, wolf, bear.

Rodentia (*n.pl.*) order of Mammalia (↑). They are herbivorous or omnivorous and gnaw their food using long incisors that grow continuously. Canine teeth are absent, e.g. squirrel, rat, mouse.

rodent (*n*) any member of the Rodentia (↑).

Primates (*n.pl.*) order of Mammalia (↑). They have a well developed brain, quick responses and binocular vision. They walk erect or semi-erect. Primates have big toes and thumbs for grasping; flexible fingers; nails are often present instead of claws. Only a few young are produced at each birth; e.g. monkey, chimpanzee, man.

Rodentia
a rodent, mouse

hominid (*n*) an animal of the family Hominidae, e.g. man and his ancestors.

Homo (*n.pl.*) genus of Primates (↑). It contains several extinct species from which *Homo sapiens* (↓) has developed.

Homo sapiens (*n*) the only existing species of man; has a large brain; prominent chin; no brow ridges; dissimilar hands and feet. *Homo sapiens* walks erect on back limbs; makes tools with hands; communicates with others of the species by speech.

human (*adj*) belonging to or showing characteristics of man.

virus (*n*) intracellular parasite of animal and plant cells. It shows no characteristics of living things outside the host cell. A virus is small enough (20–400nm in diameter) to pass through filters which retain bacteria and is only visible using an electron microscope. They are normally elongated structures in plant cells; spherical in animal cells. They consist of an outer protein shell surrounding nucleic acid, either DNA or RNA; both are never present together. Viruses cannot multiply outside living cells; they take over and use the host's cell-replication system, e.g. smallpox (↓), influenza, measles, yellow fever (↓), mumps, poliomyelitis viruses.

shapes of viruses
influenza virus
0.1 micrometre
polio virus
tobacco mosaic virus

virus

mumps virus

tobacco mosaic virus, TMV (*n*) a virus (↑) that infects tobacco plants. It is used as a model for study of plant viruses.

smallpox (*n*) disease of man caused by a virus (↑); often fatal. It has recently been eliminated from the world by a WHO eradication programme.

yellow fever (*n*) disease of man caused by a virus (↑) and transmitted by the mosquito. It is controlled by a vaccine.

interferon (*n*) protein produced by animal cells infected by viruses (↑) that protects adjacent cells from viral invasion. Protection is not limited to the virus (↑) that induced its production. It may be active also against some cancers.

bacteriophage, phage (*n*) virus (↑) which attacks bacteria. It contains DNA as the genetic material. It has a round body and a small tail. The tail attaches the virus to a bacterium, DNA of the phage enters the bacterium causing it to produce new phage; the bacterium then lyses releasing new phage.

Rickettsiae (*n.pl.*) intracellular parasites intermediate between bacteria and viruses (↑). They are similar in structure to bacteria, contain both DNA and RNA but, like viruses, can reproduce only in host cells. They are parasitic on fleas, lice, ticks, mites. If transmitted to man they cause disease, e.g. typhus.

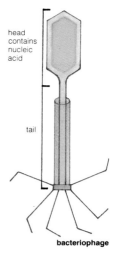

head contains nucleic acid

tail

bacteriophage

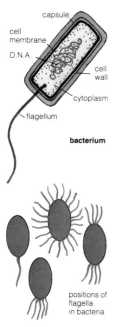

capsule
cell membrane
D.N.A.
cell wall
cytoplasm
flagellum

bacterium

positions of flagella in bacteria

binary fission of bacterium cell

bacterium (n) (pl.bacteria) prokaryote (p.13) micro-organism (0.5 to 2.0μm in diameter) visible under light microscope. A bacterial cell is surrounded by a cell wall; some have a capsule lying outside this. A cell membrane surrounds cytoplasm which contains granules but no vacuoles or organelles. Some bacteria have one or more flagella for locomotion; these are arranged in a species specific manner. Bacteria are grouped depending on shape, locomotion and on nutrients used and the products of their metabolism. Most are saprophytes or parasites. They multiply asexually by binary fission; sexual reproduction occurs sometimes. Beneficial bacteria are used in sewage disposal, cheese making; harmful ones cause disease.

aerobic bacteria (n.pl.) bacteria (↑) that need oxygen for metabolism. Contrast anaerobic bacteria (↓).

anaerobic bacteria (n.pl.) bacteria (↑) not needing oxygen for metabolism. Contrast aerobic bacteria (↑).

eubacteria (n.pl.) often called the true bacteria (↑). They have rigid cell walls; some have a flagellum; some are surrounded by a capsule (p.17). Under unfavourable conditions some can form endospores (↓).

spirochaetes (n.pl.) spiral-shaped bacteria (↑). They have flexible walls and move by flexations of the cell wall.

spirillae (n.pl.) spiral bacteria (↑). They have rigid cell walls and tufts of flagella at either end.

cocci (n.pl.) (sing.coccus) immotile, spherical bacteria (↑). They rarely have flagella. These bacterial cells often form pairs, e.g. diplococcus; chains, e.g. streptococcus; cluster, e.g.staphylococcus.

bacillus (n) (pl.bacilli) rod-shaped bacterium; may be curved or straight. Some have flagella. **bacilliform** (adj).

lactic acid bacteria (n.pl.) rod or spherical anaerobic (↑), Gram +ve (↓) bacteria that produce lactic acid; used in cheese making.

Escherichia coli, E. coli (n.pl.) Gram −ve (↓), rod-shaped bacteria (↑). They are used in the laboratory for bacterial study. They cause mild dysentery.

endospore (n) cell formed by some eubacteria (↑) to withstand unfavourable conditions, e.g. heat.

micro-organism (n) organism that can only be seen with a microscope, e.g. unicellular Algae, Protozoa, bacteria.

Gram's stain (n) stain dividing bacteria (↑) into two groups. **Gram +ve** coloured by it, **Gram −ve** not. It is named after its discoverer.

motile (*adj*) capable of movement. *Contrast immotile* (↓).
immotile (*adj*) not capable of movement. *Contrast motile* (↑).
lysozyme (*n*) enzyme which destroys bacterium by
digesting its cell wall; found in tears, egg white.
plaque (*n*) area of lysed cells; formed e.g. when phage
(p.26) destroy bacteria grown on agar culture media (↓).
incubation (*n*) the maintenance of cells, tissues, organs
or organisms under constant temperature. It is carried
out in an apparatus called an **incubator. incubate** (*v*).
inoculate (*v*) to introduce micro-organisms into, e.g. a
culture media is inoculated with bacteria.
culture (*n*) batch of micro-organisms normally grown on
a liquid culture medium (↓) in an incubator (↑). It shows
several phases; lag, exponential, stationary, decline (↓).
lag phase (*n*) period during the culture (↑) of micro-
organisms before growth starts. It occurs, e.g. in
bacteria, while they produce new enzymes to act on
substrates in the culture medium.
exponential phase (*n*) period during the culture (↑) of
micro-organisms when the number of cells increases
very rapidly. All the cells are viable (↓).
generation time (*n*) the time between two successive
divisions of a cell.
stationary phase (*n*) period after the exponential phase
(↑) in the culture of micro-organisms when the cells are
viable (↓), but not reproducing; either due to lack of
nutrients or build up of toxic metabolic products.
decline phase (*n*) period after the stationary phase (↑) in
the culture of micro-organisms when the number of
viable (↓) cells decrease.
viable (*adj*) capable of reproduction; e.g. viable bacteria
are living and can be grown in bacterial culture (↑).
culture medium (*n*) nutrient substances used to culture
(↑) micro-organisms. The medium is normally sterile to
keep the strains pure.
agar (*n*) substance extracted from sea-weed. It solidifies
the liquid culture media (↑) used for growing bacteria.
sterile (*adj*) free from viable (↑) micro-organisms.
sterilization (*n*) process of making a medium sterile (↑).
bactericidal (*adj*) describes chemicals that destroy
bacteria, e.g. phenol, and also drugs which kill them,
e.g. penicillin. *Compare bacteriostatic* (↓).
bacteriostatic (*adj*) describes chemicals and drugs,
e.g. sulphonamides, that stop bacterial growth without
killing the bacteria. *Compare bactericidal* (↑).

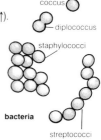

coccus
diplococcus
staphylococci
bacteria
streptococci

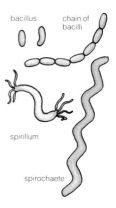

bacillus — chain of bacilli
spirillum
spirochaete

growth curve of
bacterial culture

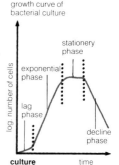

stationery phase
exponential phase
lag phase
decline phase
log number of cells
culture
time

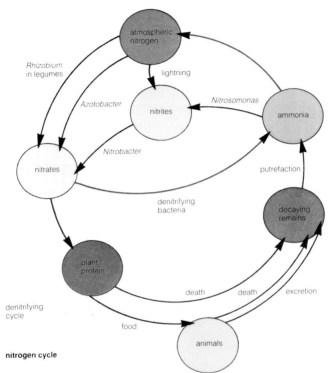

nitrogen cycle

nitrogen cycle (*n*) the circulation of nitrogen in nature.
The activities of nitrifying (p.30) and nitrogen-fixing
bacteria result in the formation of nitrates. Nitrates
are then built up into organic compounds by plants
which either decay or are eaten by animals. Nitrogen is
returned to the soil by this decay, in animal excreta or
by death and subsequent decay of the animal. It is
reconverted into nitrates by the action of *Nitrosomonas*
and *Nitrobacter* (p.30) bacteria in the soil. This
completes the cycle. Small amounts of nitrogen are lost
to the atmosphere by the action of denitrifying bacteria
(p.30); small amounts are returned to the soil by
lightning during thunder storms. In farming fertilizers
replace nitrates lost from the soil by crop cultivation.

nitrification (*n*) conversion of organic nitrogen compounds into nitrates which are used by plants. It is brought about by nitrifying bacteria, *Nitrosomonas* (↓) and *Nitrobacter* (↓). *See nitrogen cycle* (p.29).

Nitrosomonas bacteria (*n.pl.*) soil bacteria which oxidise ammonia to nitrites in the nitrogen cycle (p.29).

$$2NH_3 + 3O_2 \rightarrow 2NO^-_2 + 2H_2O + 2H^+$$

Nitrobacter bacteria (*n.pl.*) soil bacteria which oxidise nitrites to nitrates in the nitrogen cycle (p.29).

$$2NO^-_2 + O_2 \rightarrow 2NO^-_3$$

nitrogen fixation (*n*) formation of nitrogen containing compounds from atmospheric nitrogen; occurs in the nitrogen cycle (p.29). It is brought about by free-living soil bacteria, some blue-green algae and bacteria living symbiotically (↓) in root nodules (↓).

Azotobacter (*n*) genus of nitrogen-fixing bacteria (↑).

Rhizobium (*n*) genus of bacteria, found in root nodules (↓). They fix nitrogen in the nitrogen cycle (p.29).

nodule (*n*) small, rounded swelling on the root of a leguminous plant (e.g. pea, clover, bean) in which symbiotic (↓) bacteria of the genus *Rhizobium* (↑) live. They are involved in nitrogen fixation in the nitrogen cycle (p.29).

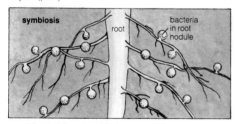

symbiosis (*n*) partnership between different kinds of organisms which benefits both, e.g. bacteria in root nodules (↑).

symbiont (*n*) an organism living in symbiosis (↑).

denitrifying bacteria (*n.pl.*) soil bacteria that act on nitrites and nitrates releasing nitrogen into the atmosphere. They cause nitrogen loss from the nitrogen cycle.

denitrification (*n*) removal of nitrogen, e.g. by denitrifying bacteria (↑) from the nitrogen cycle (p.29).

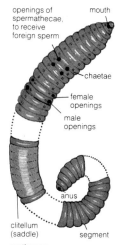

openings of spermathecae, to receive foreign sperm

mouth

chaetae

female openings

male openings

anus

clitellum (saddle)

segment

earthworm
viewed from below

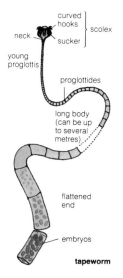

curved hooks ⎤
⎬ scolex
sucker ⎦

neck

young proglottis

proglottides

long body (can be up to several metres)

flattened end

embryos

tapeworm

worm (*n*) loose term for an elongated invertebrate animal lacking appendages. Includes helminths, platyhelminths, nematodes, annelids (p.22).

earthworm, Lumbricus (*n*) a member of the phylum Annelida. It is the common worm found in soil. Its body is surrounded by a very thin cuticle. It is composed of about 150 segments (↓), each of which, except for the first and last, have four pairs of chaetae (↓). Earthworms are hermaphrodite.

segment (*n*) unit into which the body of annelids and arthropods is divided. With minor variations each contains a similar pattern of blood vessels, nerves, excretory organs etc. Segments at the 'head' often differ.

chaeta (*n*) (*pl.chaetae*) a bristle, composed of chitin, growing from the skin of some segmented worms. It is similar to seta (p.39). Chaetae are arranged in groups on each side of the body segments (↑). In earthworms they assist in locomotion.

clitellum, saddle (*n*) area in some sexually mature annelids, e.g. leach, earthworm (↑) where the body wall is thickened. It contains glandular, mucus secreting cells which help bind two worms together during copulation. It also secretes a cocoon to hold the eggs. As the cocoon passes over the spermatheca (↓) eggs are fertilized.

spermatheca (*n*) a sac in some female and some hermaphrodite animals which receives and stores sperm until needed to fertilize the eggs. It is found in certain segments of the earthworm (↑).

worm cast (*n*) material passed out, after digestion, from the body of an earthworm.

tapeworm (*n*) parasitic flatworm (Platyhelminthes, p.22) several of which can become parasitic in man. Infestation is as a result of eating undercooked meat and fish, e.g. *Taenia* is transmitted by eating infested pork.

proglottis (*n*) (*pl.proglottides*) segment of tapeworm (↑).

scolex (*n*) 'head' of a tapeworm (↑) which attaches it to the host by **suckers** and/or hooks.

strobila (*n.pl.*) proglottides (↑) forming 'body' of a tapeworm (↑).

ascaris (*n*) parasitic roundworm (Nematoda, p.22); infests the small intestine of man, dog, fish etc. Larvae are ingested with contaminated food.

Ankylostoma, Ancylostoma (*n*) roundworm (Nematoda, p.22), causing hookworm infestation in man. Larvae penetrate skin; adult worms live in small intestine.

Fungi (*n.pl.*) group of organisms (Fungi, p.19) divided into; Phycomycetes, e.g. mildew (↓); Ascomycetes, e.g yeast (↓); Basidiomycetes, e.g. mushroom (↓), puff-ball (↓), rusts (↓), smuts (↓); Fungi Imperfecti. Fungi are simple plants made of cellular filaments called hyphae (↓). They lack chlorophyll so live either as saprophytes or as parasites. Normally they reproduce asexually by means of spores (↓) or sexually by the formation of a zygote (↓). **fungal** (*adj*).

thallus (*n*) simple vegetative plant body consisting of one or more cells. It is not differentiated into stem, root and leaves.

hypha (*n*) (*pl.hyphae*) a filament of the vegetative body or mycelium (↓) of a fungus (↑). It is composed of one or more cylindrical cells. It increases in length by growth at the tip and branches to form new hyphae.

mycelium (*n*) tangled mass of hyphae (↑) forming the vegetative body or thallus (↑) of a fungus.

haustorium (*n*) an outgrowth from a root, stem or hypha of some parasitic plants and some fungi. It penetrates the living host cell to absorb food materials.

stroma (*n*) the mass of hyphae (↑) in fungi from, or in which reproductive structures are formed.

spore (*n*) a reproductive body formed in vast numbers by plants, fungi, bacteria and protozoa. On separation from the parent it gives rise to a new individual of the species.

zoospore (*n*) motile spore formed in certain algae and fungi. It is produced in a sporangium (↓) called a **zoosporangium**. They are either ciliated or posess one or more flagella.

sporangium (*n*) (*pl.sporangia*) structure in fungi that produces asexual spores (↑).

sporangiophore (*n*) stalk which carries a sporangium (↑).

ostiole (*n*) opening in fruit bodies of some fungi (↑) for spores to escape. It is also the opening in some algae for discharge of gametes.

oogonium (*n*) simple female sex organ of some algae and fungi; contains one or more egg cells.

zygote (*n*) cell formed in plants and animals (see p.186) by the union of two gametes. In fungi it is formed by the fusion of two different mycelia. Normally it gives rise to a new individual but under unfavourable conditions it can form a zygospore (↓).

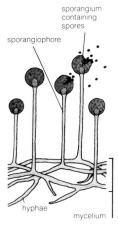

sporangium containing spores

sporangiophore

hyphae

mycelium

Fungi
a fungus, *Mucor*

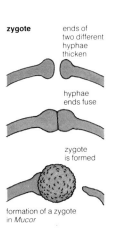

zygote ends of two different hyphae thicken

hyphae ends fuse

zygote is formed

formation of a zygote in *Mucor*

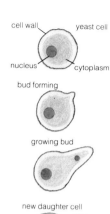

cell wall · yeast cell
nucleus · cytoplasm

bud forming

growing bud

new daughter cell

budding
budding in yeast

heterothallism (*n*) condition in algae and fungi in which sexual reproduction takes place between two separate thalli. Each thallus is self-fertile.

zygospore (*n*) a zygote (↑) which has developed a thick wall. It is a resting spore.

yeasts (*n.pl.*) unicellular fungi (↑) belonging to the Ascomycete group. They multiply asexually by budding (↓). Used in the baking and brewing industries as they can carry out fermentation, i.e. conversion of sugar into alcohol and carbon dioxide.

budding (*n*) asexual process by which unicellular plants, e.g. yeasts (↑), reproduce. A daughter cell is formed as an outgrowth of the parent cell.

mushroom (*n*) edible fruiting body of a fungus belonging to the Basidiomycetes group. The outer surface, the umbrella-shaped cap or **pileus** is supported by a stalk called a **stipe**. Inside the cap spores (↑) are formed on **gills** called **lamellae**.

spawn (*n*) the mycelium of an edible fungus like the mushroom (↑). It is used to propagate the plant.

toadstool (*n*) common name for the fruiting body of a basidiomycete fungus, other than the mushroom.

puff-ball (*n*) basidiomycete fungus similar to a mushroom (↑). It differs in that the gills are not exposed. Spores develop inside the ball which bursts when ripe.

mildew (*n*) (1) fungal disease of plants. (2) fungus causing this disease. Often synonymous with mould (↓).

mould (*n*) (1) superficial growth of fungus mycelium. (2) fungus causing this growth, e.g. mucor (↓). Often synonymous with mildew (↑).

rusts (*n.pl.*) group of basidiomycete fungi which form reddish spores. Cause damage to grain crops.

smuts (*n.pl.*) group of basidiomycete fungi which form black spores. Cause damage to grain crops.

mucor (*n*) fungus that appears as a white-grey mould on, for example, stale bread.

pythium (*n*) fungus which grows on young seedlings causing them to wilt and die.

penicillium (*n*) fungus which is visible as a blue-grey mould. It produces the **penicillins**, a group of closely related antibiotics, which, when extracted from the culture medium, kill bacteria. *Penicillium* was first observed by Alexander Fleming in 1928. Subsequent work several years later investigated its mode of action and led to its widespread use in medicine.

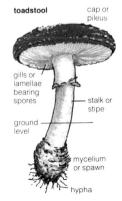

toadstool
cap or pileus
gills or lamellae bearing spores
stalk or stipe
ground level
mycelium or spawn
hypha

fish (*n.pl.*) cold-blooded (↓), aquatic vertebrate (p.23) living in salt or fresh water. External features, *see diagram*. Divided into **cartilaginous fish**, i.e. those with skeleton of cartilage (sharks and rays) and **bony fish**, i.e. those with skeleton of bone (all ordinary fish except sharks and rays).

fin (*n*) thin, flat structure projecting from the body of fish (↑). There are several types: median, dorsal, caudal, ventral, pectoral, pelvic, anal; they can be paired, unpaired, separate or continuous. Fins produce movement and control direction of movement and equilibrium.

fin ray (*n*) skeletal support for the fin (↑) of a fish.

scale (*n*) thin, protective plate-like structure on the outer surface of fish, reptiles, etc. There are several types: **cosmoid**, **ganoid**, **placoid** which vary slightly in structure and origin.

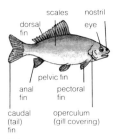

fish
external features

placoid scale

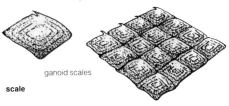

ganoid scales

scale

gill (*n*) the respiratory organ of aquatic animals. It has a large surface area and is richly supplied with blood vessels. Diffusion of oxygen and carbon dioxide between the water and the animal's vascular system takes place across it. Both internal gills (↓) and external gills (↓) occur.

internal gill (*n*) gill (↑) occurring within the gill slits (↓). They are found in most fish.

external gill (*n*) gill (↑) found on the surface in amphibians and larvae of lung fish.

gill slits (*n.pl.*) vertical openings from the pharynx to the exterior in aquatic chordates. They are paired, in a row on either side of the body. In fish and some amphibia they carry internal gills (↑) and are used in respiration. In *Amphioxus* many gill slits filter food from water passing through.

spiracle (*n*) modified first gill slit (↑) in aquatic animals.

operculum (*n*) bony cover of gill slits (↑) in fish and amphibia.

swim bladder, air bladder (*n*) gas-filled bladder in the body cavity of some fish. It allows the relative density of a fish to be matched to that of the surrounding water thus enabling it to remain easily at that depth.

clasper (*n*) projection of the pelvic fin in some fish. (e.g. sharks) to aid introduction of sperm.

flipper (*n*) limb of an aquatic mammal or bird which has become adapted for swimming.

seal

flipper

flipper ——

blubber (*n*) thick layer of fat below the skin in aquatic mammals (e.g. whale) that prevents excess heat loss.

roe (*n*) mass of eggs contained in a fish, i.e. the ovary.

spawn (*n*) mass of eggs laid in water by fish and frogs. *Compare milt* (↓).

milt (*n*) mass of sperm of a fish. *Compare spawn* (↑).

fry (*n*) a fish which has just hatched.

cold-blooded (*adj*) describes most animals except birds and mammals. They are unable to regulate body temperature by physiological means; that of aquatic animals stays close to the temperature of the surrounding water; that of terrestrial animals varies depending on air temperature, amount of radient heat from the sun, rate of muscular activity etc.

warm-blooded (*adj*) describes mammals and birds that keep their body temperature constant by physiological means, usually slightly higher than the external environment.

frog (*n*) a web-footed (↓) amphibian (p.24). Adults lack a tail and have strong hind legs for jumping. The aquatic larvae are called tadpoles (↓).

web-footed (*adj*) describes an aquatic animal or bird that has skin stretched between some digits; aids swimming.

tadpole (*n*) aquatic larva (p.36) of the frog (↑). It metamorphoses rapidly absorbing the tail and replacing the gill slits by lungs for terrestrial living.

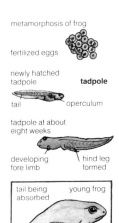

metamorphosis of frog

fertilized eggs

newly hatched tadpole

tadpole

tail   operculum

tadpole at about eight weeks

developing   hind leg fore limb   formed

tail being   young frog absorbed

frog

larva (*n*) (*pl.larvae*) immature forms of many animals after hatching from the egg and before metamorphosis (p.41) to the adult; e.g. tadpole (p.35) of frog, caterpillar of butterfly. Larval and adult forms differ structurally and in mode of life. **Larval** (*adj*).

Xenopus (*n*) genus of African aquatic toad. Formerly it was used in pregnancy testing.

carapace (*n*) protective shield of exoskeleton covering the body part of some animals, e.g. crab, turtle, tortoise.

shell (*n*) hard protective cover for the soft body of animals like crab and tortoise.

mantle (*n*) fold of skin covering most of a mollusc's body.

univalve (*n*) shell (↑) consisting of one piece or valve, e.g. found in snails.

univalve

whelk

bivalve (*n*) shell (↑) consisting of two pieces or valves hinged together, e.g. found in mussel, clam.

nekton (*n*) animals like fish and whales that swim in the pelagic zone (p.237) of a lake or sea.

bivalve

clam

plankton (*n*) very small plants, **phytoplankton** and animals, **zooplankton** that float in surface waters of seas and lakes. Smallest forms, the **nanoplankton** include diatoms; **microplankton** are a little larger. Plankton is important ecologically and economically, providing food for fish and whales. Much plankton consists of the larvae of invertebrates that are sessile (↓) when adult.

ciliary feeding (*n*) feeding method of many aquatic invertebrates. Minute organisms, e.g. plankton (↑) are filtered from water drawn into the mouth by ciliary action.

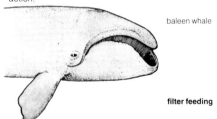

baleen whale

filter feeding

filter feeding (*n*) feeding method found in many aquatic invertebrates. Minute organisms, e.g. plankton (↑) are filtered from the surrounding water.

sessile (*adj*) describes an animal that remains in one place; it is attached to the ground or to a support.

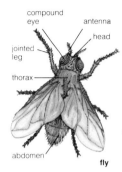

compound
eye
antenna
head
jointed
leg
thorax
abdomen
fly

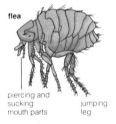

flea

piercing and
sucking
mouth parts
jumping
leg

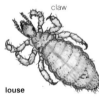

claw

louse

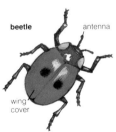

beetle
antenna

wing
cover

insect (n) a member of the class Insecta (p.23). The body is divided into head, thorax and abdomen. The head has one pair of antennae, a pair of compound eyes and mouthparts for feeding. The thorax has three pairs of jointed legs and usually two pairs of wings. Many insects are terrestrial; they have waxy cuticles to resist dry conditions and breathe air through trachea. Others are aquatic. Most have a metamorphosis in their life cycle. Some are pests, e.g. locust (↓) others are parasites, e.g. flea (↓) many are vectors (p.211) of disease.

flea (n) wingless, blood-sucking insect (↑), parasitic on mammals and birds. The body is flattened; legs are well developed; mouthparts are adapted for piercing and blood-sucking. They transmit disease, e.g. typhus.

louse (n) (pl.lice) wingless insect (↑) parasitic on mammals. It has a flat body, short legs with claws for attachment to the host and mouthparts adapted for blood-sucking. Eggs of lice, called **nits**, are attached to the host's hair. They transmit disease, e.g. typhus.

beetle (n) a land or freshwater insect (↑). Forewings are reduced to horny covers for the membranous hind wings which may be small or absent. Some are pests, e.g. larvae of woodworm; others beneficial, e.g. **ladybirds**.

fly (n) (pl.flies) an insect (↑) (e.g. fruit fly, housefly) with only one pair of wings. Hind wings are modified to form balancing organs. Some have mouthparts adapted to suck blood or plant juices. Blood-suckers may spread disease, e.g. mosquito (malaria).

locust (n) migratory, terrestrial, winged insect (↑). It has narrow, hardened forewings, membranous hind wings and enlarged legs for jumping. Locusts sometimes form swarms and then destroy vegetation over large areas.

butterfly (n) insect (↑) of the order Lepidoptera (p.23). It has two pairs of large wings. Both wings and body are covered with scales which are often brightly coloured. The adult has a proboscis for sucking plant nectar. It has a complete metamorphosis in the life cycle. Larvae, or caterpillars (p.38), feed mainly on leaves of plants. Butterflies differ from moths (↓) in being active in daylight, having clubbed antennae and resting with wings folded over the back.

moth (n) insect (↑) of the order Lepidoptera (p.23). Moths differ from butterflies (↑) in being mostly nocturnal, often less brightly coloured, never having clubbed antennae, and resting with their wings in various positions.

caterpillar (n) larva or grub (p.42) of a butterfly, moth or certain other insects. They have soft, worm-like bodies with three pairs of legs on the thorax and prolegs on the abdomen.

ant (n) small, social insect (↓). It has an ovipositor (p.40). Adults have two pairs of membranous wings hooked together for stable flight.

termite, white ant (n) a social insect (↓). They are pests of timber. Their nests are mainly underground but they can form huge ant hills above ground.

bee (n) social insect (↓). It has a sting; both biting and sucking mouthparts; a thorax joined to the abdomen by a narrow waist; two pairs of membranous wings in adults which are hooked together for stable flight.

honey bees have three castes (↓): **queens** are fertile females; **drones** are males; **workers** are sterile females that do all the work of the colony.

hive, beehive (n) case or box for keeping bees (↑).

honey (n) thick, sugary substance made by bees (↑) from the nectar of flowers.

social insects (n.pl.) a group of insects (p.37) living and working together in a highly organised community. They are divided into castes (↓).

caste (n) type of insect living in a colony, e.g. bees (↑), wasps, ants (↑). Distinction between castes in bees and ants is on a sexual basis; in termites it is not sex related.

swarm (n) dense mass of insects of the same species, e.g. a group of bees going off to form a new colony.

bug (n) loose term for certain insects, especially one that infests houses and bedding; properly applied to sucking insects of the order Hemiptera, e.g. aphids.

tick (n) an arachnid (p.23). It is a blood-sucking parasite of man and animals. It can transmit disease, e.g. typhus.

mite (n) an arachnid (p.23). There are several species, most parasitic on specific animals. They can transmit disease, e.g. typhus.

scorpion (n) an arachnid (p.23) that lives in hot climates. Its head and thorax are united. The segmented abdomen ends in a narrow tail with a sting used in defence.

spider (n) an arachnid (p.23). It has pointed appendages for catching and sometimes poisoning prey; a head and thorax united and joined to the abdomen by a narrow waist; an abdomen carrying spinnerets that produce silk for trapping prey, building egg cocoons, lining nests.

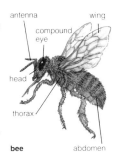

antenna · wing · compound eye · head · thorax · abdomen

bee

tick

mite

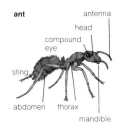

ant — antenna, head, compound eye, sting, abdomen, thorax, mandible

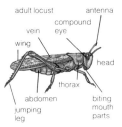

adult locust — antenna, compound eye, vein, wing, head, thorax, abdomen, biting mouth parts, jumping leg

tiger moth — antenna, compound eye, thorax, abdomen, wing

tagma (*n*) (*pl.tagmata*) one of the three main divisions into which the body of arthropods are divided, i.e. head, thorax (↓), abdomen (↓).

thorax (*n*) in insects the tagma (↑) behind the head. It consists of three segments and carries the three pairs of legs and, if present, the wings.

abdomen (*n*) posterior tagma (↑) of arthropods, the segments of which are similar to each other.

cephalothorax (*n*) the fused head and thorax (↑) of some crustaceans and arachnids.

wing (*n*) membrane structure carried on the thorax (↑) of an insect. It has veins (↓) and is used for flying.

vein (*n*) one of the fine tubes of toughened cuticle forming the framework of an insect wing (↑).

venation (*n*) arrangement of veins in an insect wing (↑). It is used in classification as it is species specific.

appendage (*n*) projection from the body, e.g. arms of vertebrates, antennae (↓) of insects, pincers (↓) of crab.

antennae (*n.pl.*) (*s.antenna*) a pair of appendages (↑) on the head of many arthropods. They may appear club shaped or feathery. They always occur in pairs and are much jointed; used for touch and smell.

antennule (*n*) one of a pair of appendages (↑) on the head of a crustacean; used for touch and smell. Crustaceans also have a pair of antennae (↑).

feeler (*n*) loose term for a tentacle (↓) or antenna (↑).

tentacle (*n*) slender, flexible, non-jointed organ on the head of an invertebrate. Usually more than one are present, e.g. an octopus has eight. Tentacles may be used for feeling, attachment, gripping prey or swimming.

cercus (*n*) (*pl.cerci*) a tail-like appendage (↑) at the hind end of many arthropods, especially insects. They are always paired. Modified in earwigs to form pincers (↓).

pincer, chela (*n*) (*pl.chelae*) a claw with a pincer-like movement for picking up objects. They are found on certain appendages (↑) of some crustaceans and arachnids.

clasper (*n*) outgrowth from some animals, particularly insects where it occurs on one of the rear abdominal segments. It is used to grip the female during copulation.

seta (*n*) bristle-like outgrowth or chitin scale on the cuticle of an arthropod.

haltere (*n*) one of a pair of organs found in flies of the order Diptera (p.23). They are much reduced, modified hind wings.

ovipositor (*n*) a tube-like organ at the hind end of the abdomen (p.39) of female insects. Eggs are laid through it and can be positioned in an inaccessible place. The sting of a worker bee is a modified ovipositor.

spinneret (*n*) one of several (usually six) organs found at the end of the abdomen (p.39) in spiders. Ducts from the spinning gland (↓) open on to it.

spinning gland (*n*) a gland found in caterpillars and spiders. It produces liquid silk which hardens in the air. The silk is used to make spiders' webs (↓), trap prey, line nests and make the cocoons of caterpillars.

web (*n*) (1) fine mesh woven by a spider from the liquid silk produced in the spinning glands (↑); used to trap prey; (2) the skin of webbed feet (p.35).

proboscis (*n*) a long, tube-like structure that projects from the head of an animal, e.g. the trunk of an elephant; the sucking mouthparts (↓) of an insect.

mouthparts (*n.pl.*) appendages (p.39) on the head of an arthropod. They are used for feeding and often modified to suit a particular method.

mandible (*n*) mouthpart (↑) of an arthropod. The edge is usually toothed for cutting and crushing solid food.

maxilla (*n*) (1) mouthpart (↑) of an arthropod lying behind mandibles (↑). (2) bone in upper jaw of vertebrates which carries all upper teeth except incisors.

labrum (*n*) upper lip of an insect.

labium (*n*) lower lip of an insect.

fat body (*n*) (1) network of tissue under the skin and round the gut of insects. It stores reserve food which is used before metamorphosis (↓) and during hibernation. (2) in amphibians and lizards it is one of two masses of fatty tissue found in the abdomen. It is used as a food reserve during hibernation.

simple eye, ocellus (*n*) simple light receptor found in insects and some other invertebrates.

compound eye (*n*) type of eye found in insects and crustaceans. It consists of numerous, separate units each of which has a lens and light-sensitive cells and can form an image of an object. The units are separated from each other by pigmented cells. Each unit forms an image of a small portion of the object being viewed. The whole is seen as a combination of all the small images. There are differences in the method of formation of the image in the eyes of diurnal and nocturnal insects.

web
common garden spider and web

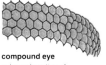

compound eye
enlarged section of compound eye of insect

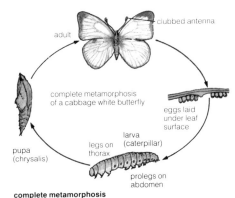

complete metamorphosis
of a cabbage white butterfly

clubbed antenna

adult

eggs laid
under leaf
surface

larva
(caterpillar)

legs on
thorax

pupa
(chrysalis)

prolegs on
abdomen

complete metamorphosis

metamorphosis
complete metamorphosis
of a house fly

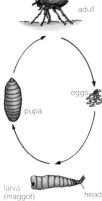

adult

eggs

pupa

larva
(maggot)   head

metamorphosis (*n*) stage in the life history of some
animals involving rapid transformation from the larval to
the adult form, e.g. caterpillar to butterfly, tadpole to
frog. It often involves destruction of tissues and the
formation of new ones and is controlled by hormones. In
an insect **complete metamorphosis** includes larva,
pupa, and imago (p.42) stages. This occurs, e.g. in
housefly and butterfly. **Incomplete metamorphosis**
involves nymph (p.42) and adult stages only. This
occurs, e.g. in grasshopper. **Metamorphic** (*adj*).

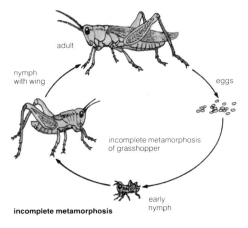

adult

nymph
with wing

eggs

incomplete metamorphosis
of grasshopper

incomplete metamorphosis

early
nymph

maggot (*n*) legless, worm-like insect larva.

grub (*n*) an insect larva (p.36).

cocoon (*n*) a protective cover for eggs and developing
larvae and pupae produced by many invertebrates,
e.g. the citellum of an earthworm produces a cocoon to
contain the eggs; some spiders spin a cocoon to hold
their eggs, larvae of many insects spin a cocoon in
which the pupae (↓) develop. The fabric silk is made
from the cocoons of silkworms.

cocoon

silk moth pupa

pupa

cocoon

pupa (*n*) (*pl.pupae*) resting stage between larva (p.36)
and adult in the life cycle of insects that undergo
complete metamorphosis (p.41). Feeding and
movement stop while large changes in the structure
occur which alter larvae into adults. Pupae are often
encased in cocoons (↑). **Pupal** (*adj*). **Pupate** (*v*).

chrysalis (*n*) the pupal (↑) stage of butterflies, moths and
certain other insects.

nymph (*n*) young, immature stage in the life cycle of
insects that undergo incomplete metamorphosis (p.41).
It has compound eyes and similar mouthparts to the
adult but differs in being sexually immature and either
lacks wings or has them only partially developed.
Nymphs that are adapted to an aquatic habitat are
called **naiads** but the term nymph is often applied to
them also.

imago (*n*) a sexually mature insect. It is the final adult
stage of both incomplete and complete metamorphosis
(p.41).

ecdysis (*n*) moulting(↓). In arthropods it is periodic and
involves the shedding of the cuticle of immature
individuals; it allows growth to occur. In reptiles (except
crocodiles) it occurs throughout life and involves
periodic shedding of the outer layer of epidermis.

instax (*n*) stage of an insect larva between two ecdyses
(↑).

moult (*n*) (1) synonymous with ecdysis (↑). (2) the
shedding of hair or feathers by birds or mammals. It
takes place periodically, depending in some cases on
growth rate and on seasonal changes in others.

beak

body covered
with feathers

wing

foot

tail

claw

bird
external features
of a typical bird

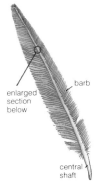

enlarged
section
below

barb

central
shaft

feather
structure of a feather

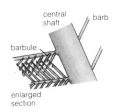

central
shaft

barb

barbule

enlarged
section

bird (*n*) general term for a feathered animal; most can fly. A vertebrate of the class Aves (p.24).

wing (*n*) the forelimb of a bird (↑) which is adapted for flight. It is covered with feathers.

adapt (*v*) to change or modify a generalised feature for a particular purpose, e.g. the forelimbs of birds are adapted for flight but use of the resulting wings is restricted to flying; webbed feet are adapted for swimming; plant tendrils enable it to cling to a support.

feathers (*n.pl.*) structures that cover the body of birds (↑). Like hair they consist of dead keratinized cells and can be moved by muscles. Feathers are moulted (↑) once a year. A feather consists of a central shaft, barbs (↓) and barbules (↓); different arrangement of these gives rise to three types of feather; contour feather (↓); down feather (↓); filoplume (↓).

contour feathers (*n.pl.*) large feathers of a bird (↑) which overlap and cover the body, wings and tail. They are arranged in rows and give the bird its shape.

down feathers (*n.pl.*) small, soft feathers of a bird (↑). In adult birds they are found under the contour feathers (↑) and cover the body. In young chicks (↓) they are the only feathers present.

chick (*n*) the young of some birds, particularly the hen.

filoplumes (*n.pl.*) small, hair-like feathers of birds (↑) found between the contour feathers (↑) on the body.

quill feather (*n*) a contour feather (↑) found on the wings or tail. It is used for flying.

plume (*n*) alternative term for one or a tuft of feathers.

plumule (*n*) a small feather (↑) or a small plume (↑), e.g. a down feather (↑).

plumage (*n*) the colouring and arrangement of bird feather.

preen gland (*n*) an oil gland in a bird. The oil it secretes is used to preen (↓) feathers.

preen (*v*) the action in birds of applying oil from the preen gland (↑). It is carried out by the beak. At the same time the contour feathers (↑) are smoothed and arranged.

quill (*n*) (1) the hollow portion of the shaft of a feather (↑). (2) the whole shaft of a quill feather (↑). (3) a hollow spine on a porcupine.

barb (*n*) a delicate, thread-like structure on the shaft of a feather. It forms the web of the feather.

barbule (*n*) a small, hooked process on the barb (↑) of a feather. Barbules interlock with those on other barbs.

beak (*n*) (1) horny, pointed bill on the head of a bird; variously shaped depending on the type of food eaten. (2) similar shaped projections in other animals or plants.

animal (*n*) a living, organised being: a member of the animal kingdom. It respires, excretes, increases in size, reproduces. In contrast to plants (↓) it moves from place to place, cannot make its own food from simple molecules and shows quicker responses to the environment.

biped (*n*) animal which uses a single pair of limbs for locomotion, e.g. man.

quadruped (*n*) animal, usually a mammal, which uses two pairs of limbs for locomotion, e.g. cow, horse.

tetrapod (*n*) animal with two pairs of limbs. Included are amphibians, reptiles, birds, mammals.

hoof (*n*) (*pl. hooves*) modified nail (p.80) of some animals, e.g. cow.

ungulate (*n*) animal with hooves (↑). They are herbivores, and live in herds on hard, open ground, e.g. antelope.

horn (*n*) (1) hard outgrowths on head of an animal, e.g. cattle, giraffe, rhinoceros; also the tentacle of a snail. (2) hard substance, keratin, of which horns are made.

antler (*n*) a branched horn (↑) found in some mammals, e.g. deer. They are shed every year.

claw (*n*) sharp, pointed, curved projection at the end of a digit, limb or appendage in animals, birds, insects.

pad (*n*) a protective area of thick skin under the foot or toes of some vertebrates, e.g. dog.

talon (*n*) a large, sharp, hooked claw found in some birds of prey, e.g. eagle, owl.

nest (*n*) shelter made by animals, birds, insects and some fish, e.g. newt, to hold eggs until they hatch (↓) and/or young until they can look after themselves.

incubation (*n*) the maintenance of a developing egg at a controlled temperature. Heat is supplied from the body of a parent, the sun or artificially. **Incubate** (*v*).

incubation period (*n*) time between egg laying and hatching.

hatch (*v*) to emerge from the egg as a fully formed animal, e.g. a nestling breaks out of its shell.

crop (*n*) in birds, a sac in the alimentary canal to store food temporarily. In invertebrates it both stores and digests food.

air sac (*n*) an air-filled sac in birds and insects which helps respiration during flight.

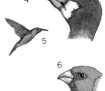
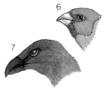

beak

beak adaptations for various diets

1. Bunting – eats corn and other seeds. Beak strong and wedge shaped
2. Great tit – eats insects and tree buds. Beak very short and strong
3. Duck – eats weeds, grain, small aquatic animals. Beak can sift mud and water
4. Cormorant – fish eater. Long beak to catch live prey
5. Hummingbird – eats insects and nectar. Long beak to reach into flowers
6. Hawfinch – eats large seeds and fruit stones. Large, strong beak for crushing hard outer covering
7. Vulture – flesh eater. Strong, hooked beak for tearing carrion

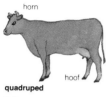

horn

hoof

quadruped
cow

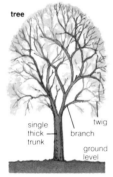

tree

single
thick
trunk

twig

branch

ground
level

**woody
perennials**

shrub

several
thinner
branches

twig

ground
level

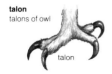

talon
talons of owl

talon

plant (*n*) an organism belonging to the plant kingdom. Plants respire, increase in size and reproduce. Except for fungi they make their own food by photosynthesis. In contrast to animals (↑) they cannot move from place to place and make slow responses to the environment.

vegetation (*n*) the mass of plants present in an area.

vegetable (*n*) whole or part of a plant, other than fruits, used as a food, e.g. carrot, rice, pulses.

annual (1) (*adj*) something which occurs every year. (2) (*n*) a plant which completes its life cycle (i.e. from germination of the seed to production of a new seed) in one growing season. It lives for one year, e.g. maize.

biennial (1) (*adj*) something occurring in alternate years. (2) (*n*) a plant which takes two growing seasons to complete its life cycle (i.e. from germination of the seed to production of a new seed). During the first season it stores food to be used during flowering and seed production in the second. It lives two years, e.g. wallflower.

perennial (1) (*adj*) something occurring every year. (2) (*n*) a plant that takes many years to reach full size and begin to reproduce. Two types: **herbaceous perennials**, where non-woody aerial parts die away in autumn to be replaced the following spring by new growths from underground organs; **woody perennials** where woody stems survive winter above ground and growth continues from them next spring, e.g. trees.

shrub (*n*) a low (up to 4m) woody, perennial (↑) plant. Unlike a tree (p.59) it lacks a single, thick trunk but has several thinner branches formed from near the ground.

herb (*n*) (1) a plant without a woody stem or parts which survive from year to year above ground. (2) a plant used in medicine and cookery.

geophyte (*n*) a plant that survives winter by means of underground buds, e.g. bulbs.

vascular plant (*n*) plant with a vascular system of xylem to conduct water, and give mechanical support; and phloem to conduct food. Includes ferns, gymnosperms, angiosperms.

epiphyte (*n*) plant growing on and supported by another plant, e.g. moss. It does not get food from the support.

liana, liane (*n*) a climbing plant, especially the woody ones that grow rapidly in tropical forests.

succulent (1) (*n*) a plant that stores water in its tissues, e.g. cactus. (2) (*adj*) describes a plant, stem, leaf or fruit which is full of sap or juice.

taxis, tactic movement (*n*) movement of an organism from place to place in response to a stimulus. Positive taxis is towards the stimulus; negative taxis away from it. Can be preceded by a prefix describing the nature of the stimulus, e.g. **chemotaxis** is movement in response to chemicals, **thermotaxis** to temperature, **thigmotaxis** to touch. *Compare tropism, nasty* (↓).

phototaxis (*n*) tactic (↑) movement in response to light.

geotaxis (*n*) tactic movement (↑) in response to gravity.

hydrotaxis (*n*) tactic movement (↑) in response to water.

tropism, tropic movement (*n*) growth curvature of part of a plant in response to a stimulus. Direction of curvature depends on direction of stimulus; positive tropism is towards the stimulus, negative tropism is away from it. If preceded by a prefix describes nature of the stimulus, e.g. phototropism, geotropism, hydro-tropism, chemotropism, thermotropism, helio-tropism, photoperiodism (↓). *Compare taxis* (↑), *nasty* (↓).

phototropism (*n*) tropic (↑) response to light. Stems of plants are positively phototropic.

geotropism (*n*) tropic (↑) response to gravity; roots of plants are positively geotropic and grow downwards; stems, negatively geotropic, grow upwards.

hydrotropism (*n*) tropic (↑) response to water. Plant roots grow towards water.

chemotropism (*n*) tropic (↑) response to chemicals.

thermotropism (*n*) tropic (↑) response to temperature variation.

heliotropism (*n*) tropic (↑) response to sunlight.

photoperiodism (*n*) response in plants and animals to the relative length of day and night.

nasty, nastic movement (*n*) response made by plants to a stimulus. Unlike tropism (↑) it does not depend on the direction of stimulus. Prefix, if added, describes nature of stimulus. They may be slow growth curvatures, e.g. the opening and closing of some flowers in response to light intensity, i.e. **photonasty** or rapid changes caused by changes in turgor, e.g. folding and drooping of leaflets of mimosa when touched, i.e. **thigmonasty**.

nyctinasty, sleep-movements (*n.pl.*) nastic (↑) move-ments in response to daily changes of light and/or temp-erature; usually close at night, open in the day, e.g. marigold. Some (e.g. evening primrose) open at night.

autonomic movements (*n.pl.*) movements in plants produced by internal stimuli, e.g. unfolding of buds.

phototropism
positive phototropic
response of plants

light from one side only

light from all sides

geotropism
geotropic response
of germinating
bean seed

negative geotropism – shoot grows upwards

positive geotropism – root grows downwards

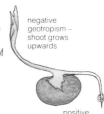

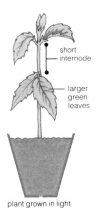

plant grown in light

- short internode
- larger green leaves

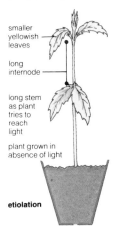

etiolation

- smaller yellowish leaves
- long internode
- long stem as plant tries to reach light

plant grown in absence of light

variegation

variegated leaf of ivy

cream parts lack chlorophyll

transpiration (*n*) loss of water vapour by diffusion from the surface of land plants. It occurs mostly during the day, mainly through stomata of leaves and to a slight extent through the cuticle. Excess is harmful, causing wilting (↓). It is the driving force behind movement of water in xylem vessels from roots and is thus ultimately responsible for turgor (p.12) and transport of mineral salts.

root pressure (*n*) pressure that results in passage of water from root cells into xylem vessels of plants; mechanism unknown.

potometer (*n*) instrument to measure rate of passage of water into a cut or intact plant.

wilt (*v*) to become flaccid (↓) through lack of water. Plant cells loose turgidity (↓), hence, in herbaceous plants, mechanical support and the plant droops.

flaccid (*adj*) describes cell tissue which is soft, non-rigid.

turgid (*adj*) describes a cell in which entry of water by osmosis has caused maximum cell volume.

etiolation (*n*) condition of green plants grown in dark. Plants lack chlorophyll hence have small, yellow leaves; internodes are lengthened resulting in long thin stems.

variegation (*n*) differential, irregular colouration of plant parts; inherited or caused by disease.

gland (*n*) in plants, small organ secreting chemicals – gum (↓) tannin (↓), latex (↓), resin (↓) – to the exterior, e.g. hairs of stinging nettle, nectaries in many flowers.

gum (*n*) plant secretion that is sticky when liquid but hard when dry, e.g. gum arabic. A gum forming a slimy solution in water is called **mucilage**.

latex (*n*) liquid secretion exuding from cut surfaces of some flowering plants. Usually milky, e.g. secretion of rubber tree, dandelion.

resin (*n*) acidic, organic secretion produced by many trees and shrubs, particularly conifers.

tannin (*n*) colourless, organic plant secretion used in tanning and dying.

alkaloids (*n.pl.*) basic, nitrogen-containing, organic compounds. Some are poisonous, e.g. strychnine; many are used in medicine, e.g. morphine, cocaine, quinine.

auxin (*n*) plant hormone controlling growth; produced by cells in growing apices of stems and roots; uneven distribution causes tropic movements (↑).

essential elements (*n.pl.*) elements needed in relatively large amounts by plants for nutrition. Those needed in very small amounts are **trace elements**, see p.122.

epidermis (*n*) in plants the outer layer of tissue. Usually it consists of one layer of rectangular cells that fit closely together to protect the plant. Aerial epidermis is often covered by protective cuticle.

piliferous layer (*n*) the region of the epidermis (↑) in plant roots that bears root hairs (↓).

exodermis (*n*) layer of protective cells found under the epidermis (↑) of roots. It becomes the outer layer in older roots where the piliferous layer (↑) has withered.

endodermis (*n*) a layer of rectangular, close-fitting cells in plants. It is the innermost layer of the cortex surrounding the vascular tissue; controls transfer of water and dissolved substances between them. Found in all roots, and stems of ferns and some dicotyledons.

root (*n*) usually underground part of a plant that grows downwards, anchoring the plant in ground. It absorbs water and mineral salts from soil. Roots differ from stems in internal structure, do not bear buds or leaves but may have root hairs (↓) and root caps (↓).

root cap (*n*) loose mass of cells covering the tip of a growing point in a root; gives protection.

root hair (*n*) outgrowth from an epidermal cell; found in large numbers in the piliferous layer (↑) of roots. They increase the surface area of roots. Nutrients and water pass from the soil across their thin walls.

ground tissue (*n*) the connective tissue of a plant; includes pith (↓), cortex (↓), medullary rays (↓), hypodermis (↓). It usually consists of parenchyma (↓) but can contain sclerenchyma (↓) and collenchyma (p.50).

pith, medulla (*n*) central core of a stem that has cylindrical vascular tissue. It is usually made up of parenchyma cells. It stores food. Pith is also found in roots where central tissue develops into parenchyma rather than xylem. **Hollow pith**: *see diagram.*

cortex (*n*) area of mainly parenchyma tissue found between the epidermis (↑) and endodermis of roots and stems in vascular plants. It can store nutrients, e.g. starch.

medullary ray (*n*) one of a large number of thin, vertical plates of parenchyma cells running radially through the vascular tissue from the inside to the outside of a stem. It conducts and stores food material.

hypodermis (*n*) layer of cells under the epidermis (↑) of some plant stems and leaves. It can be mechanically strengthened to give extra protection (e.g. pine) or it can store water (e.g. succulent leaves).

root
root of runner
bean seed

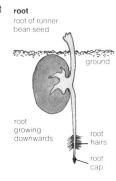

ground

root
growing
downwards

root
hairs

root
cap

root

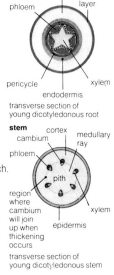

cortex   piliferous
          layer
phloem

pericycle   xylem
          endodermis
transverse section of
young dicotyledonous root

stem   cortex   medullary
cambium   ray
phloem

          pith

region
where   xylem
cambium
will join   epidermis
up when
thickening
occurs

transverse section of
young dicotyledonous stem

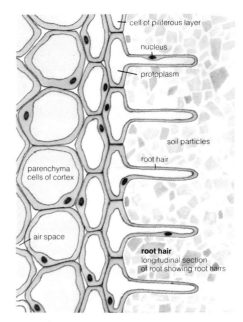

cell of piliferous layer

nucleus

protoplasm

soil particles

root hair

parenchyma cells of cortex

air space

root hair
longitudinal section of root showing root hairs

parenchyma (*n*) in plants a tissue composed of living, thin-walled cells which are round ended and almost as broad as they are long. Air spaces occur between the cells. Found in pith (↑), cortex (↑), mesophyll (p.50). It forms the ground tissue of plants and is the basic tissue from which others, e.g. collenchyma (p.50) and sclerenchyma (↓) are formed.

sclerenchyma (*n*) plant tissue consisting of thick-walled cells which provide mechanical support; usually have no living protoplasm when mature. Two types of cell are found, fibre cells (↓) and stone cells (↓).

fibre cell (*n*) very long, tapering cell found in sclerenchyma (↑). Fibre cells occur singly or in strands, e.g. flax. They can withstand mechanical strain.

bast (*n*) type of fibre cell (↑) found in sclerenchyma (↑).

stone cell, sclereid (*n*) cell found in sclerenchyma (↑), not much longer than it is broad. Stone cells occur in leaves, fruit, seed coats and support surrounding tissue.

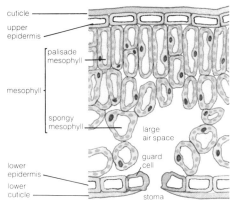

cuticle

upper epidermis

palisade mesophyll

mesophyll

spongy mesophyll

large air space

guard cell

lower epidermis

lower cuticle

stoma

transverse section of a leaf

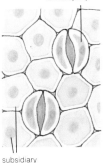

stoma
surface of epidermis of dicotyledon leaf showing stomata open and closed

guard cell (turgid therefore stoma open)

subsidiary cells

guard cell (flaccid therefore stoma closed)

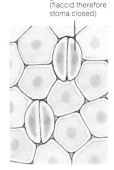

collenchyma (*n*) plant tissue found under the epidermis (p.48) which provides mechanical support for growing stems, leaves, flowers. Its elongated cells are thickened with cellulose mainly at the corners so that they can stretch as the surrounding cells grow.

mesophyll (*n*) tissue between the upper and lower epidermis (p.48) of a leaf blade; consists mainly of parenchyma cells. It is differentiated into palisade mesophyll (↓) and spongy mesophyll (↓).

palisade mesophyll, – layer (*n*) mesophyll (↑) found below the upper epidermis of a leaf. Cells are slightly elongated and arranged perpendicularly to the leaf surface in 2–3 layers. It contains many chloroplasts and is particularly concerned in photosynthesis.

spongy mesophyll, – layer (*n*) mesophyll (↑) found above the lower epidermis of a leaf; consists of irregular, loosely arranged cells which contain few chloroplasts. Large air spaces, found between the cells, communicate with the outside air through stomata. It is particularly concerned with transpiration and gaseous exchange.

meristem (*n*) tissue in plants formed of cells which undergo cell division. **Apical meristems** are found at the tips of stems and roots in flowering plants and ferns, **secondary meristems** arise from older permanent tissues, e.g. cork cambium.

cambium (*n*) meristem (↑) found between xylem and phloem of vascular bundles. It is responsible for the increase in thickness of a stem.

vascular bundle
vascular bundles arranged in a random manner

transverse section of monocotyledon stem

ring-like system of vascular bundles

transverse section of dicotyledon stem

vascular cylinder
pericycle
piliferous layer
primary xylem
primary phloem
endodermis

arrangement of vascular tissue in root systems

stoma (*n*) (*pl.stomata*) (1) one of a large number of small pores or openings in the epidermis of aerial parts of plants, particularly the underside of leaves. Through them exchange of gases takes place and water is lost by transpiration. Each is surrounded by a pair of guard cells (↓) (2) includes pore and guard cell (↓).

guard cell (*n*) one of a pair of crescent-shaped epidermal cells that surround a stoma (↑) in plants. Changes of turgor alter their shape causing opening and closing of the pore. Stomata thus control transpiration rate and rate of gaseous exchange.

subsidiary cell (*n*) plant epidermal cell surrounding a guard cell (↑). Helps guard cell open and close stoma (↑).

stele, vascular cylinder (*n*) central cylinder of vascular tissue in stem and roots of vascular plants; surrounded by endodermis. It consists of xylem (p.52), phloem (↓), pericycle (p.52) and in some cases also pith and medullary rays. Arrangement varies in different plants.

vascular system (*n*) tissue, mainly xylem (p.52) and phloem (↓) forming a continuous series of vessels in all parts of vascular plants. Conducts water, mineral salts, synthesized food and provides mechanical support.

vascular tissue (*n*) plant tissue, mainly xylem (p.52) to conduct water and phloem (↓) to conduct food; also some sclerenchyma, for strengthening and parenchyma.

vascular bundle (*n*) longitudinal strand of vascular tissue (↑) in stems and roots of vascular plants. Consists of xylem and phloem with some supporting tissue. It is a unit of the stele (↑). Can be arranged in a ring around the pith as in gymnosperms and dicotyledons or randomly in stem tissues as in monocotyledons and ferns.

phloem (*n*) vascular tissue (↑) of plants which conducts synthesized food. Contains sieve tubes (↓) and in some plants companion cells (↓), fibres (p.15) and parenchyma cells. **primary phloem** is formed first, **secondary phloem** is found where secondary thickening has occurred.

sieve tube (*n*) conducting, tubular element of phloem (↑) in plants. Consists of thin-walled, elongated cells that are connected through the end walls by means of groups of perforations called **sieve plates**.

callose (*n*) carbohydrate which is deposited on sieve plates (↑) to seal them off either permanently or seasonally. Produced naturally or in response to injury.

companion cell (*n*) small specialised parenchyma cell associated with sieve tubes (↑) of flowering plants.

xylem (n) part of the vascular tissue (p.51) of plants. Contains tubes to conduct water and dissolved minerals from roots to leaves; also provides mechanical support. Consists of two main types, tracheids (↓) and vessels (↓) plus fibres and parenchyma. **Primary xylem** is formed first, **secondary xylem** is formed when secondary thickening occurs. Most of the stem and roots of trees is xylem and is called **wood**.

tracheid (n) elongated; empty, dead cell in xylem of vascular plants. It runs parallel to the long axis. Walls are thick, lignified and pitted. Its long ends overlap adjacent cells with which it communicates via pits (p.15). Conducts water; provides mechanical support.

xylem vessels (n.pl.) found only in ferns and angiosperms. Formed of xylem cells joined end to end by **perforation plates**, i.e. where cell wall material is removed allowing free passage of sap. A vertical series of xylem vessels is a **trachea**.

pericycle (n) thin layer of parenchyma cells, sometimes including fibres, surrounding the stele (p.51) and within the endodermis. Found in gymnosperms and dicotyledons. Its cells may become meristematic and are the origin of branch roots in dicotyledons.

translocation (n) movement of substances in vascular plants by conducting tissues, e.g. xylem (↑) and phloem.

cuticle (n) in plants a non-cellular layer of cutin secreted by the epidermis. Over the aerial parts it is broken only by stomata (p.51). Prevents excess water loss.

cork (n) protective layer of dead, empty cells found on the outside of woody plants, under scale leaf scars and at the site of wounds. Formed by cork cambium. It replaces the epidermis of young stems and roots.

bark (n) protective tissue of dead cells found on the outside of stems and roots of woody plants. It may consist of cork (↑) only or alternating layers of cork and dead cortex; some dead phloem cells may also be present. Commonly all tissues outside the xylem.

lenticel (n) small pore, usually eliptical, in the bark of a woody stem that allows gaseous exchange between the air and the interior of the stem when the epidermis has been replaces by cork layers.

passage cell (n) cell in the endodermis of some plant roots. It remains unthickened during secondary growth to allow passage of water and dissolved substances between the cortex and the vascular cylinder.

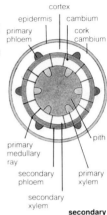

secondary thickening

transverse section of young dicotyledonous stem

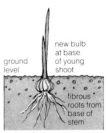

root system
root system of young onion plant, a monocotyledon

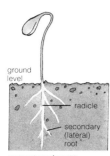

root system of young
sunflower plant, a
dicotyledon

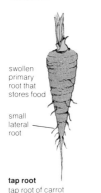

swollen
primary
root that
stores food

small
lateral
root

tap root
tap root of carrot

adventitous root
adventitous roots on runner
of strawberry

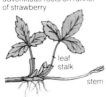

leaf
stalk

stem

adventitous roots
at node of the stem

root system (n) the arrangement of roots in plants. There are two types, the tap root (↓) system of dicotyledons and the fibrous system (↓) of monocotyledons.

radicle (n) root of the embryo in seed plants. The first structure of the germinating seedling to appear; often emerges from the micropyle.

primary root (n) the plant root which develops from the radicle (↑) of the seed.

lateral root, secondary root (n) a root which is smaller than the main root and branches outwards from it.

rootlet (n) the smallest, final branch of a root.

tap root (n) root system where the primary root remains predominant, though very much smaller lateral roots may be present, e.g. turnip, carrot.

adventitious root (n) root which has developed from a part of the plant other than the root, i.e. from a stem or leaf cutting, e.g. in tomato, strawberry, grasses.

fibrous root (n) one of a tuft of adventitious roots (↑) in plants. They grow from the base of a stem or hypocotyl; are of approximately equal diameter; normally bear smaller lateral roots (↑), e.g. wheat, corn, onion.

prop root (n) an adventitious root (↑) which grows from a node on the branch of a tree. It grows downwards into the soil and provides support for the branch.

buttress root, stilt root (n) one of several adventitious roots (↑) that grow from one or two nodes near the ground on the main plant stem. They grow downwards into the soil and provide support.

aerial root (n) adventitious root (↑) that grows from a node on the stem of a plant. It does not grow downward into the soil. Found, e.g. in orchids where they absorb water from the air and in ivy, where, as the plant matures they root in the ground and function as normal roots.

root tuber (n) swollen, underground root that stores food and is the organ of vegetative reproduction, e.g. dahlia.

pneumatophore, breathing root (n) root that arises as a lateral branch of a tap root system (↑). It grows upwards into the air and supplies the root system with oxygen. Found in mangrove trees that grow in swamps.

rhizoid (n) hair-like structure that serves as a root. Found, e.g. at the base of the stem in moss, on the under-surface of a fern prothallus and in liverwort.

holdfast, hapteron (n) a disc at the end of either a thallus or a stalk to attach a plant to its support.

stem (*n*) the aerial parts of a vascular plant. It has a growing point or points at the terminal end. Stems may have leaves and buds in regular patterns at nodes (↓) and reproductive structures, e.g. flowers. Stems translocate food and water, provide support and may store food. They can be modified, e.g. cladode (↓), thorn (↓), tendril (↓) or can form runners, stolons and offsets for vegetative propagation (p.77). Some stems may be underground organs for overwintering, e.g. rhizomes, corms, tubers; these still have the characteristics of stems, i.e. leaves with buds in their axils and vascular bundles arranged in a ring or randomly.

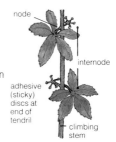

Virginia creeper   **tendril**

node (*n*) region of a stem where one or more leaves are attached.

internode (*n*) region of a stem between two adjacent nodes. It does not have leaves.

stalk (*n*) a long structure, usually thinner than a stem (↑). It provides support, e.g. leaf stalk.

shoot (*n*) the stem of a young vascular plant that develops from the plumule. It bears leaves.

creeper (*n*) herbaceous plant with a stem that grows along the ground putting down roots along its length or climbs to obtain more light by twining its climbing stem (↓) round a support.

climbing stem (*n*) stem (↑) that climbs upwards by using other structures for support.

twining stem (*n*) stem (↑) that twists round and round a support. Plants of one species twine in the same direction.

cactus
cladode

rootstock (*n*) upright, underground stem (↑). The upper end is at ground level, at which point it commonly bears a basal rosette of leaves, e.g. dandelion.

tendril (*n*) stem, leaf or part of a leaf in climbing plants which is modified to form a thin thread-like structure. It either twines round a support, e.g. pea, grape, or sticks to it, e.g. virginia creeper.

rose
hook

cladode (*n*) modified stem that resembles a leaf in appearance and function, including photosynthesis, e.g. butcher's broom.

thorn (*n*) short, pointed, hard structure on a plant. Formed from a side shoot that has lost its apical growing point.

hook (*n*) curved structure in a plant. Morphologically similar to a thorn (↑).

blackthorn
spine

spine (*n*) a sharp, rigid structure in a plant. It is a modified leaf or stipule.

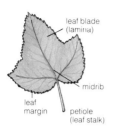

simple leaf
simple leaf of vine

simple leaf
examples of common
leaf shapes

lanceolate

sagittate

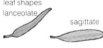

compound leaf

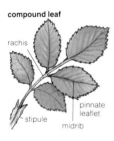

rachis

pinnate
leaflet

stipule

midrib

palmate leaf

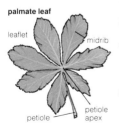

leaflet

midrib

petiole

petiole
apex

prickle (*n*) a sharp, pointed outgrowth from the epidermis of the stem (↑) in plants.

barb (*n*) a stiff, hooked hair in plants.

foliage (*n*) the mass of non-specialised leaves (↓)
  foliage leaves on the aerial parts of a plant, usually a tree or shrub.

leaf mosaic (*n*) pattern formed by the arrangement of leaves on stems or branches such that they do not completely overlap and so obtain maximum sunlight.

leaf (*n*) an outgrowth at the node of a stem in plants. It consists of a leaf base, a stalk or petiole (p.56) and a thin, flattened portion, the leaf blade or lamina (p.56). It is generally green and the lamina contains mesophyll cells which carry out photosynthesis. Leaves are also involved in transpiration. They produce lateral buds in their axils. Most plants produce several types of leaf including seed leaves or cotyledons, protective scale leaves, floral leaves, i.e. petals, sepals and foliage (↑) leaves. Leaves may be simple (↓) or compound (↓).

hastate   cordate   ovate   pinnatisect

simple leaf (*n*) leaf (↑) that has the leaf blade in one piece. The margin may be indented but not as far as the midrib. Different names are given to the various shapes.

compound leaf (*n*) leaf (↑) that has the leaf blade divided into separate segments or leaflets (↓). The margins extend to the midrib.

leaflet (*n*) unit or segment of a compound leaf (↑).

pinna (*n*) a leaflet (↑) on a pinnate leaf (↓).

pinnate leaf (*n*) a compound leaf with leaflets arranged on either side of a rachis (↓). There may or may not be a terminal leaflet.

rachis (*n*) the axis of a pinnate, compound leaf (↑) to which leaflets (↑) are attached.

palmate leaf (*n*) a compound leaf with separate leaflets (↑) attached to the apex of the petiole.

deciduous (*adj*) describes a perennial plant, usually a tree or shrub, that sheds its leaves at a particular season, i.e. autumn in temperate climates. It passes through a period without leaves; e.g. oak, beech, horse-chestnut.

evergreen (*adj*) describes a perennial plant, usually a tree or shrub, that has leaves all the year round. Leaves are constantly shed and replaced; e.g. holly, conifers.

phyllotaxy, phyllotaxis (*n*) the arrangement of leaves
  on a stem or axis in plants, e.g. alternate (↓), decussate
  (↓), opposite, spiral or whorled (↓). Phyllotaxy partially
  determines the leaf mosaic (p.55).
alternate (*adj*) describes phyllotaxis (↑) in which one leaf
  grows at each node, successive leaves grow on
  opposite sides of the stem.
decussate (*adj*) describes phyllotaxis (↑) in which a pair
  of leaves at each node are opposite each other and
  are set at right angles to the pairs of leaves above
  and below. *See leaf scars of horse-chestnut twig
  (p.57)*.
whorled (*adj*) describes phyllotaxis (↑) where more than
  two leaves grow from one node.
bract (*n*) small, leaf-like structure in plants with a
  relatively undeveloped blade. It is usually green. Flower
  and inflorescence stalks arise in its axil.
bracteole (*n*) small bract (↑) on a flower stalk or at the
  base of a flower in an inflorescence.
bract scale (*n*) one of several scales on the female cone
  of coniferous plants, e.g. pine, fir.
needle (*n*) long, stiff leaf found, for example, in pine.
frond (*n*) (1) leaf-like structure of a fern. It has the same
  functions as a leaf. (2) the leaf of a palm tree.
leaf base (*n*) structure in plants which attaches the leaf
  to a stem.
leaf stalk, petiole (*n*) stalk which supports the lamina (↓)
  of a leaf in plants.
sessile (*adj*) in plants, describes a leaf without a petiole
  (↑).
leaf sheath (*n*) a modification of the base of a leaf in
  plants. It forms a sheath around the stem. Found in
  grasses and other monocotyledons.
ligule (*n*) thin, dry membrane extending above the leaf
  sheath (↑) at the junction of the lamina in a sessile (↑)
  leaf. Found particularly in grasses and sedges.
stipule (*n*) one of two processes found on either side of
  the petiole in many plants. It protects the axillary bud
  (↓). It may be leaf-like and carry out photosynthesis or it
  may be in the form of a tendril or a spine.
leaf blade, lamina (*n*) the flattened part of a leaf in
  plants. There are various shapes; each plant species
  has a characteristic shape. The arrangement of
  vascular tissue is called venation (↓). Vascular tissues
  of leaf blades are primary xylem and phloem.

parallel venation
parallel veins of
maize, a monocotyledon

leaf margin
examples of leaf margins

entire

serrate

dentate

spiny

sinuate

net venation
net veins of
nasturtium, a dicotyledon

leaf scar
horse chestnut twig

leaf margin (*n*) the edge of a leaf which may be a different colour to the rest of the leaf blade (↑). Leaf margins have various shapes, *see diagram*.

vein (*n*) one of the vascular bundles in a leaf blade (↑) of a plant. Consists of primary xylem and phloem. Arrangement of veins is called venation (↓).

midrib (*n*) the large central vein of a leaf blade in plants.

venation (*n*) (1) the veins (↑) of a leaf blade. (2) the arrangement of these veins which may be parallel (↓) or reticulate (↓).

parallel venation (*n*) arrangement of leaf veins (↑) found in monocotyledonous plants. There are several parallel veins of approximately equal size.

reticulate venation, net venation (*n*) arrangement of leaf veins (↑) found in dicotyledonous plants. A network of small veins branch from larger primary veins.

axil (*n*) the angle in plants between the upper side of a leaf and the stem from which it arises. Lateral buds (↓) and axillary buds (↓) are formed here.

axillary bud (*n*) bud found in the axil (↑) of a leaf in plants.

lateral bud (*n*) bud found in the axil (↑) of a leaf in plants. It develops if the apical bud (↓) is damaged.

apical bud (*n*) terminal bud at the apex of a stem or branch in plants. It is responsible for increase in height.

bud (*n*) undeveloped shoot, normally found in the axils of leaves at the apex of a stem in plants. Also an unopened flower or an unopened leaf.

scale leaf (*n*) a strong leaf in a plant, e.g. bud scales which protect buds.

apical dominance (*n*) the inhibition of the growth of lateral buds by the apical buds in herbaceous plants and trees. If the terminal bud is damaged the upper lateral buds start to develop and continue the upward plant growth.

abscission (*n*) process by which two parts of an organ are separated, e.g. a leaf from a stem.

abscission layer (*n*) a layer of parenchyma cells at the base of the leaf stalk in woody dicotyledons. When these cells break down the leaf is shed. A layer of cork plugs the vascular tissue. A leaf scar (↓) is left on the stem. Also found in flower stalks and fruits.

leaf fall (*n*) loss of leaves after abscission (↑) in perennial plants and trees. Two types occur, one in deciduous and the other in evergreen trees.

leaf scar(*n*) scar on plant that marks where the leaf was attached to the stem before abscission (↑).

photosynthesis (n) a process in green plants in which complex organic compounds are synthesised from water and carbon dioxide using energy absorbed from sunlight by chlorophyll (↓). The reaction takes place in chloroplasts (↓) in green plants and in chromatophores (↓) in blue-green algae. The overall reaction is:

$$6CO_2 + 12H_2O \rightarrow C_2H_{12}O_6 + 6O_2 + 6H_2O$$

The products are glucose (which is normally converted to starch), molecular oxygen and water. Photosynthesis is directly or indirectly the source of carbon and energy for all forms of life. The initial stage of the process is a light reaction (↓); this is followed by a dark reaction (↓).

light reaction (n) the first reaction in photosynthesis (↑). It is light dependent, requires chlorophyll and results in the splitting of water to yield oxygen, reducing power and energy. The energy and the reducing power are used in the subsequent dark reaction (↓).

dark reaction (n) the second reaction in photosynthesis (↑). It is not dependent on light. The energy and reducing power supplied by the light reaction (↑) are used to convert carbon dioxide to carbohydrate.

photosynthetic (adj) describes organisms that obtain energy from sunlight.

photosynthetic pigments (n) the light-absorbing pigments of photosynthesis (↑). Included are chlorophylls (↓), carotenes (↓) and xanthophylls (↓).

pigment (n) a substance which colours animal or plant tissues.

chlorophyll (n) green pigment found in all plants except fungi and a few flowering plants. Chlorophylls are complex organic compounds that take part in the light reaction (↑) of photosynthesis. Those found in green plants are chlorophylls a and b; only chlorophyll a is found in blue-green algae. Except for blue-green algae, chlorophyll occurs in chloroplasts.

bacteriochlorophyll (n) the main photosynthetic pigment found in photosynthetic bacteria.

carotene (n) orange or yellow photosynthetic pigment found in chloroplasts, chromatophores and plastids, e.g. carrot root. Vertebrate liver changes carotene into vitamin A.

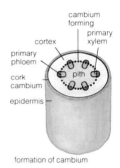

formation of cambium

secondary thickening
in dicotyledonous stem
(transverse section)

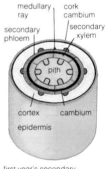

first year's secondary
thickening

xanthophyll (*n*) yellow photosynthetic pigment; usually occurs with carotenc. It gives deciduous leaves their colour at leaf fall.

chloroplast (*n*) a plastid (p.14) containing chlorophyll (↑). Photosynthesis is carried out in it. Found in many cells of leaves and young stems of plants except blue-green algae.

chromoplast (*n*) a plastid (p.14) found in plant cells. Contains pigment other than chorophyll, usually orange or red. It is responsible for the colour of fruits, e.g. tomatoes and other plant parts, e.g. carrot root.

chromatophore (*n*) (1) a chromoplast (↑) in plants. (2) in bacteria and blue-green algae, an organelle containing photosynthetic pigments.

tree (*n*) a tall, woody perennial plant, usually above 3m in height. Normally it has a single trunk. In contrast to shrubs (p.45) it only forms branches some distance from the ground. It contains heartwood (↓) and sapwood (↓).

sapling (*n*) a young tree.

trunk (*n*) The main stem of a tree which does not normally have branches near the ground, *see p.45*.

branch (*n*) any structure which arises from a larger structure, e.g. the branch of a tree which grows from the trunk (↑), *see diagram p.45*.

twig (*n*) a small branch (↑) or shoot, particularly of a tree or shrub, *see diagram p.45*.

wood (*n*) (1) a group of growing trees. (2) trees which have been felled and cut into pieces for commercial use. (3) the hard part of trees and shrubs, i.e. the secondary xylem. It is the major part of their stem and roots.

sapwood (*n*) the outer layer of pale, soft, young wood in the trunk or branch of a tree. It surrounds the heartwood (↓). Xylem vessels in it conduct water and dissolved mineral salts and also provide mechanical support.

heartwood (*n*) the dark, hard, central old wood of a tree trunk. It lies within the sapwood (↑). Its cells are non-living and the xylem is non-conducting. It only functions in support. The cell walls are impregnated with lignin, tannins and resins that give heartwood its darker colour and make it more resistant to decay than sapwood.

sap (*n*) liquid in a plant, usually that which contains dissolved nutrients. It passes through the conducting vessels, i.e. xylem and phloem.

hardwood (*n*) wood (p.59) from dicotyledonous plants. It contains tracheids, parenchyma, fibres and vessels. *Compare softwood* (↓).

softwood (*n*) wood (p.59) from conifers. It contains tracheids and parenchyma. *Compare hardwood* (↑).

primary growth (*n*) growth that gives rise to a plant consisting of primary tissues (i.e. primary xylem and primary phloem) formed by the activity of apical meristems.

secondary thickening, – growth (*n*) the formation of extra supporting tissue and extra vascular tissue by the activity of the cambium. Found in dicotyledon and gymnosperm plants. It occurs only in plants where aerial stems do not die down seasonally. It results in an increase in thickening of the stem and root. The first year's secondary thickening of a dicotyledonous stem and that of a dicotyledonous root are shown in the diagrams.

annual ring, growth ring (*n*) the annual increase in the amount of secondary xylem, i.e. wood in stems and roots of woody plants in temperate climates. The secondary xylem formed in spring has large wood elements and that formed in autumn has small wood elements. This is visible in a cross-section of a trunk or stem as a series of concentric circles. It is not seen in tropical climates because growth is uniform throughout the year. The number of annual rings gives the age of the tree.

annular thickening (*n*) type of secondary wall formation or thickening of the cell wall in primary xylem vessels and tracheids of plants. Thickening is by laying down rings of lignin at intervals along the length of the cell. This allows the cell to be stretched lengthwise as surrounding cells grow. *Compare spiral, reticulate, scalariform, pitted thickening* (↓).

spiral thickening (*n*) type of thickening of the wall in xylem vessels and tracheids of plants. Thickening is by a spiral or coil of lignin round the inner surface of the cell wall. *Compare annular* (↑), *reticulate, scalariform, pitted thickening* (↓).

reticulate thickening (*n*) type of thickening of the wall in xylem vessels and tracheids of plants. Thickening is by a network of lignin over the inner surface of the cell wall. *Compare annular, spiral* (↑), *scalariform, pitted thickening* (↓).

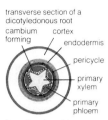

transverse section of a dicotyledonous root

formation of cambium

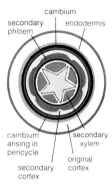

secondary thickening
first year's secondary thickening

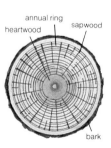

annual ring

adventitous bud which can give rise to a new plant

adventitous
adventitous leaf buds in *Bryophyllum*

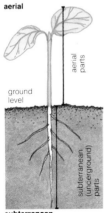

aerial

aerial parts

ground level

subterranean (underground) parts

subterranean
germinating castor oil seed

scalariform thickening (*n*) type of thickening of the wall in xylem vessels and tracheids of plants. Thickening is by a series of bars of lignin on the inner surface of the cell wall. The pattern of bars is like rungs of a ladder. Compare *annular, spiral, reticulate* (↑), *pitted thickening* (↓).

pitted thickening (*n*) type of thickening of the wall in xylem vessels and tracheids of plants. Thickening is by deposition of lignin over the whole inner surface of the cell wall except for many small pits or pores. *Compare annular, spiral, reticulate, scalariform thickening* (↑).

centripetal (*adj*) in plants, describes something showing development from outside inwards such that the newest structure is internal, e.g. phloem.

centrifugal (*adj*) in plants, describes something showing development from the inside outwards such that the newest structure is external, e.g. xylem of a stem.

adventitious (*adj*) describes tissues, organs or structures arising in abnormal positions, e.g. plant roots that develop from nodes in stems; buds that do not develop from leaf axils but from roots or from leaves.

caducous (*adj*) describes a structure that is not persistent, e.g. sepals which fall as a flower opens.

aerial (*adj*) describes something which is in the air, e.g. stem and leaves of a plant which are above ground. *Contrast subterranean* (↓).

subterranean, underground (*adj*) describes something lying below ground level, e.g. plant roots. *Contrast aerial* (↑).

vegetative (*adj*) describes whole or parts of an organism concerned with producing growth of the organism, e.g. plant stems, roots, leaves. Also describes parts used in non-sexual reproduction, e.g. bulbs, corms, suckers. Does not include parts concerned with sexual reproduction, e.g. flowers.

callus (*n*) (1) tissue that forms over a wound on the surface of woody plants. (2) plugs of callose that block sieve plates of sieve tubes (p.51).

lignification (*n*) the deposition of lignin in cell walls of sclerenchyma, xylem vessels and tracheids. It stiffens cell walls and thus provides support.

nutation (*n*) spiral course of growth shown by the apex of, for example, plant stem, root, flower stalk, tendril. Caused by continuing circular change in position of the fastest growth region of the apex.

actinomorphic (*adj*) flower with regular shape and radial symmetry, e.g. buttercup which can be bisected vertically in two or more planes into identical halves.

zygomorphic (*adj*) flower with one plane of symmetry, e.g. sweet pea. When bisected along the plane of symmetry it forms two halves which are mirror images.

flower (*n*) specialised reproductive stem found in angiosperms. Consists of an axis or receptacle (↓) from which arise sepals, petals, stamens, carpels. Petals and sepals are not directly concerned with reproduction and so are called **accessory flower parts**. Carpels and stamens are concerned with reproduction and are the **essential flower parts**. **floral** (*adj*). **flower** (*v*).

blossom, bloom (*n*) a single flower (↑); also the state of being in flower, i.e. when the flowers are open.

floret (*n*) (1) a small flower, e.g. in grasses. (2) a small single flower which when grouped with other florets forms a composite flower.

floral diagram (*n*) a diagram in the form of a ground plan showing the relative position and number of sepals, petals, stamens and carpels of a flower. A floral diagram and a diagram of half a flower are both required to describe a flower fully. The information from a floral diagram can be summarised in a floral formula (↓).

floral formula (*n*) a method of summarising the information given in a floral diagram (↑) using numbers and letters. The calyx is represented by K and followed by the number of sepals, e.g. K5 indicates five free sepals, K(5) indicates there are five sepals joined together. The corolla is represented by C, e.g. C5, C(5). The perianth (i.e. sepals and petals not distinguishable from each other) is represented by P, e.g. P6 or P(6). The androecium is represented by A, e.g. A5 indicates five free anthers, A(5) indicates they are united. A∞ represents numerous free anthers, A(∞) numerous joined ones. Two types of anther are shown as A(5+5). The gynoecium is represented by G followed by the number of carpels, e.g. G5, G(5). A superior gynoecium is shown as G$\underline{5}$, an inferior one as G$\overline{5}$. Examples: the floral formula of a buttercup is K5, C5, A∞, G$\underline{\infty}$, i.e. it has 5 free sepals, 5 free carpels, numerous free stamens and numerous free, superior carpels; the sweet pea which has 5 joined sepals, 5 joined petals, 10 stamens of two different types and one superior carpel has a floral formula K(5), C(5), A(5+5), G1.

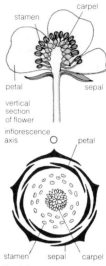

floral diagram

floral formula: K5, C5, A∞, G$\underline{\infty}$

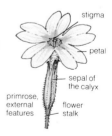

flower
snapdragon,
external features

flower
primrose,
external
features

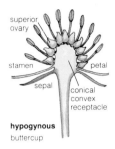

superior ovary

stamen

petal

sepal

conical convex receptacle

hypogynous
buttercup

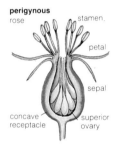

perigynous
rose

stamen

petal

sepal

concave receptacle

superior ovary

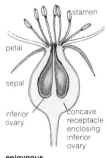

stamen

petal

sepal

inferior ovary

concave receptacle enclosing inferior ovary

epigynous
apple

longitudinal section
of types of
floral receptacle

flower stalk, pedicel (*n*) the stem which supports a single flower or an inflorescence.

peduncle (*n*) the stalk of an inflorescence. It can also be called a pedicel (↑).

rachis (*n*) the main axis of an inflorescence, *see diagrams* (p.68, 69) or of a fern frond.

receptacle, torus, thalamus (*n*) the upper end of the flower stalk in flowering plants. Its shape can vary from conical to concave. From it arise sepals (↓), petals (p.64), stamens and carpels (p.64). The carpels can be superior or inferior (p.65) to the receptacle.

hypogynous (*adj*) describes a flower in which the carpels are at the apex of a conical receptacle; the sepals (↓), petals (p.64) and stamens (p.64) are borne on the receptacle at a point below the carpels (p.64). The gynoecium is superior, e.g. buttercup.

perigynous (*adj*) describes a flower in which the carpels (p.64) are at the centre of a concave receptacle; sepals (↓), petals (p.64) and stamens (p.64) are borne round its margin.

epigynous (*adj*) describes a flower in which the receptacle completely encloses the carpels; sepals (↓), petals (p.64) and stamens (p.64) are borne above the carpels (p.64). The gynoecium is inferior, e.g. dandelion, apple.

perianth (*n*) the outer part of a flower within which lie stamens and carpels (p.64). In monocotyledons it usually consists of two whorls (↓) which are similar. In dicotyledons they are dissimilar. The outer whorl, the calyx (↓) consists of green sepals (↓) and the inner whorl, the corolla (↓) is composed of coloured petals (p.64).

whorl (*n*) a circle of, e.g. leaves, petals (p.64), sepals (↓) which arise from the stem of a plant at one level.

calyx (*n*) the outermost part of a flower which surrounds the corolla (↓). It consists of a whorl (↑) of leaf-like sepals (↓).

epicalyx (*n*) a whorl (↑) of bracts or bracteoles found outside the calyx (↑) in some flowers.

sepals (*n*) the individual parts of the calyx (↑) of a flower. They are usually green. Sepals enclose and protect the other flower parts when the flower is in bud.

corolla (*n*) the coloured petals (p.64) of a flower. It lies within the calyx (↑) and surrounds the stamens (p.64) and carpels (p.64). The lower parts of the petals may be joined to form a **corolla tube**.

petals (*n*) the individual parts of the corolla (p.63) of a flower. Usually brightly coloured and conspicuous. The shape, colour and arrangement of petals is adapted to the method of pollination, e.g. in wind-pollinated flowers petals are small or absent, in flowers pollinated by insects the petals are usually brightly coloured, large and flat.

snapdragon longitudinal section

nectar

nectary (*n*) a gland that secretes nectar (↓). Found in flowers, particularly those pollinated by insects. It is located in such a position that an insect sucking nectar will bring about cross-pollination by rubbing its body against the stamens and stigmas.

nectar (*n*) a sugary substance secreted by the nectaries (↑) of a flower. It attracts, for example, insects and humming birds which bring about cross-pollination by rubbing the body against the stamens and stigmas.

nectar (*n*) a sugary substance secreted by the nectaries (↑) of a flower. It attracts, for example, insects and humming birds which bring about cross-pollination while collecting nectar.

androecium (*n*) collective name for the male reproductive parts of a flower; these are the stamens (↓).

stamen (*n*) the male reproductive organ of a flower. It consists of a stalk or filament (↓) with an anther (↓) at its apex. It produces pollen. Collectively stamens are called the androecium (↑).

filament (*n*) part of the stamen (↑) in a flowering plant. It is stalk-like and supports the anther (↓).

anther (*n*) part of the stamen (↑) in a flowering plant. It is borne at the apex of the filament (↑). It consists of two joined lobes. Inside each lobe lie two **pollen sacs**; they produce pollen. When ripe, anthers split to release pollen.

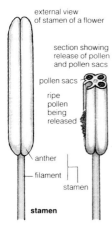

external view of stamen of a flower

section showing release of pollen and pollen sacs

pollen sacs

ripe pollen being released

anther

filament

stamen

stamen

gynoecium (*n*) collective name for the female reproductive organs or carpels (↓) of a flower. The gynoecium of a single flower may consist of one or more carpels. Each carpel is made up of an ovary (↓) and a stigma (↓). The latter is often borne on a style (↓).

carpel (*n*) a female reproductive organ in a flowering plant. Collectively they make up the gynoecium (↑). There may be one or several carpels present in each flower; flowers with one carpel are monocarpous; those with many are polycarpous. Carpels may be free (i.e. **apocarpous**) or several may be joined to form a single structure (i.e. **syncarpous**). Each carpel consists of an ovary (↓) and a stigma (↓). The latter is often borne on a style (↓).

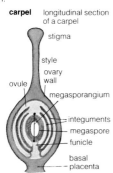

carpel longitudinal section of a carpel

stigma

style

ovary wall

ovule

megasporangium

integuments

megaspore

funicle

basal placenta

pistil (*n*) (1) alternative name for a gynoecium (↑). (2) a single carpel of a gynoecium (↑) that has separate carpels.

ovary (*n*) in flowering plants the hollow region at the base of a carpel (↑). A simple ovary can contain one or more ovules (↓). Several ovaries can be joined together to form a compound ovary.

superior ovary (*n*) ovary (↑) in a hypogynous or perigynous (p.63) flower. Carpels are either at the apex of a conical receptacle or at the centre of a concave receptacle.

inferior ovary (*n*) ovary (↑) in an epigynous (p.63) flower. Carpels are completely enclosed by and fused with the receptacle.

marginal placenta

stigma (*n*) the terminal portion of a carpel (↑) in flowering plants. Pollen grains stick to it when the flower is pollinated. Stigmas are usually, but not necessarily, borne on a style (↓).

style (*n*) an elongation of the carpel in a flowering plant. It bears the stigma (↑) and carries it into a prominent position to receive pollen. Styles are not present on all carpels.

axile placenta

ovule (*n*) structure found in conifers and flowering plants. After fertilization it develops into the seed. The ovule is unprotected in conifers; in flowering plants one or more ovules are enclosed in and protected by a carpel (↑). They are attached to the ovary wall at the placenta (↓) by a stalk called a funicle (↓). Ovules are surrounded by a protective integument (↓).

parietal placenta

funicle (*n*) stalk in flowering plants that attaches an ovule (↑) to the area of ovary wall called the placenta (↓).

integument (*n*) protective layer surrounding the ovule (↑) of seed plants. Most gymnosperms have one integument, angiosperms usually have two. After fertilization they form the seed coat.

free central placenta

placenta (*n*) part of the ovary (↑) in a plant to which ovules (↑) are attached. There are various ways of arranging placentas; this is called placentation (↓).

ovule

placentation (*n*) the arrangement of placentas (↑) in a plant ovary. Gynoecia consisting of either one carpel or several free carpels can show, for example, marginal placentation as in pea, basal placentation as in sunflower. Gynoecia formed from several joined carpels show, for example, parietal placentation as in violet, axile as in lily, free central as in primrose.

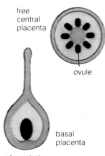

basal placenta

placentation

alternation of generations (*n*) condition shown by many plants in which the life cycle shows an alternation of a haploid generation with sexual reproduction (gameto-phyte ↓) and a diploid generation showing asexual repro-duction (sporophyte ↓). These generations often differ in appearance. In flowering plants the gametophytes are produced from megaspores (↓) and microspores (↓); they are small and have no free existence. The sporophyte generation is very much larger being the plant itself. Alternation of generations is also shown by many animals, especially coelenterates, but here there is no change in chromosome number.

sporophyte (*n*) diploid stage in life cycle of plants that show alternation of generations (↑). During it asexual spores are produced by meiosis from which gametophyte (↓) stage develops.

megasporangium (*n*) sporangium in plants that produces megaspores (↓). It is the ovule of flowers.

microsporangium (*n*) sporangium in plants that produces microspores (↓). It is the pollen sac of flowers.

megaspore (*n*) larger of the two spores produced by plants that have two kinds of spore; they give rise to the female gametophyte generation. In flowering plants it is the embryo sac. *Compare microspore* (↓).

microspore (*n*) smaller of the two spores produced by plants that have two kinds of spore; they give rise to the male gametophyte generation. In flowering plants they are the pollen grains. *Compare megaspore* (↑).

pollen (*n*) mature microspores (↑) produced in pollen sacs of an anther in seed plants. When released it is carried, usually by wind or insects, to ovules of gymnosperms or stigmas of angiosperm carpels. Here pollen grains germinate and produce pollen tubes to carry male gametes to the ovule to fertilize female gametes.

pollen tube (*n*) tube formed by a germinating pollen grain in plants. Carries male gametes to the female gamete.

gametophyte (*n*) haploid stage in life cycle of plants that show alternation of generations (↑) during which sex cells or gametes (↓) are produced. Union of male and female gametes produces the sporophyte generation (↑).

gametangium (*n*) plant organ that produces gametes.

gamete, germ cell (*n*) a reproductive cell. In plants, union of male and female gametes produced from pollen grains and embryo sac respectively gives rise to a sporophyte generation.

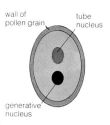

wall of pollen grain

tube nucleus

generative nucleus

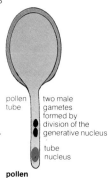

pollen tube

two male gametes formed by division of the generative nucleus

tube nucleus

pollen
development of a pollen grain

pollination (*n*) transfer of pollen from an anther to a stigma in flowering plants and from a male to a female cone in conifers. Both cross-pollination (↓) and self-pollination (↓) occur. Transfer of pollen is mainly by wind and insects, sometimes by water. After pollination the pollen germinates, produces a pollen tube and fertilization (↓) occurs.

self-pollination (*n*) a type of inbreeding in flowering plants; results if pollen is transferred from an anther to a stigma of the same flower or another flower on the same plant. Many flowers have a mechanism to prevent it occurring; some allow it only if cross-pollination (↓) fails.

cross-pollination (*n*) a type of outbreeding in flowering plants; results if pollen is transferred from an anther of a flower on one plant to the stigma of a flower on another plant of the same species. Unlike self-pollination (↑) it needs a vector, e.g. wind, insects, to carry it out.

fertilization (*n*) in plants, a process of sexual reproduction occurring after pollination (↑), resulting in the fusion of two gametes. It is brought about when a pollen tube grows from the pollen grain to bring the male gamete close to the female gamete within the ovule; fusion then occurs.

self-fertilization (*n*) fertilization (↑) of a female gamete by a male gamete from the same individual. *Compare cross-fertilization* (↓).

cross-fertilization (*n*) fertilization (↑) of a female gamete by a male gamete from a different individual of the same species. *Compare self-fertilization* (↑).

sterility (*n*) the inability to reproduce sexually.

self-sterility (*n*) inability of a hermaphrodite plant or animal to produce viable offspring by self-fertilization (↑). Viable offspring only result after cross-fertilization (↑).

incompatibility (*n*) mechanism in flowering plants to prevent self-fertilization. Usually result of a failure of a pollen tube to grow down the style after pollination.

pin-eyed (*adj*) describes a flower with a long style so that the stigma is at the mouth of the corolla tube; stamens are short. Presence of both it and thrum-eyed (↓) flowers of the same species, e.g. primrose, aids cross-pollination by insects.

thrum-eyed (*adj*) describes a flower with long stamens which reach the mouth of the corolla tube; style is short. Presence of both it and pin-eyed (↑) flowers of the same species, e.g. primrose, aids cross-pollination by insects.

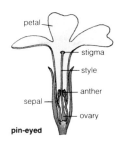

petal
stigma
style
anther
sepal
ovary
pin-eyed

longitudinal sections
of primrose

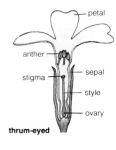

petal
anther
sepal
stigma
style
ovary
thrum-eyed

monoecious (*adj*) describes (1) plant species with separate male and female flowers on the same plant, e.g. hazel; (2) hermaphrodite animals, e.g. earthworm, that produce both male and female gametes in the same individual.

dioecius (*adj*) describes (1) plant species with male and female flowers on separate plants, e.g. willow; (2) animals that produce male and female gametes in separate animals, e.g. man.

staminate (*adj*) describes male flowers that have stamens but no carpels. *Compare pistillate* (↓).

pistillate (*adj*) describes female flowers that have carpels but no stamens. *Compare staminate* (↑).

protandrous (*adj*) describes (1) flowers in which stamens ripen before the stigma is mature; prevents self-fertilization (p.67); (2) hermaphrodite animals that produce sperm before they produce eggs. *Compare protogynous* (↓).

protogynous (*adj*) describes (1) flowers in which stigmas mature before stamens ripen. (2) hermaphrodite animals that produce eggs before they produce sperm. *Compare protandrous* (↑).

monanthus (*adj*) describes a plant with only one flower.

polyanthus (*adj*) describes a plant with several flowers.

polypetalous (*adj*) describes a flower in which petals are not joined, e.g. buttercup. *Compare sympetalous* (↓).

sympetalous, gamopetalous (*adj*) describes a flower in which petals are joined, e.g. primrose. *Compare polypetalous* (↑).

monopetalous (*adj*) describes flowers, either those with one petal or those with joined petals, i.e. sympetalous (↑).

polysepalous (*adj*) describes a flower in which the sepals are not joined, e.g. buttercup. *Compare synsepalous* (↓).

synsepalous, gamosepalous (*adj*) describes a flower with joined sepals, e.g. primrose. *Compare polysepalous* (↑).

monosepalous (*adj*) describes either flowers with one sepal or those with joined sepals, i.e. synsepalous (↑).

monocarpous (*adj*) describes flowers with only one carpel.

polycarpous (*adj*) describes flowers with more than one carpel.

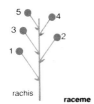

rachis

raceme

simple raceme, eg. lupin

rachis

panicle

panicle, eg. oat

racemose inflorescences
numbers show order of flower opening

rachis

corymb
corymb, eg. candytuft

spike

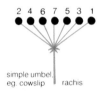

spike,
eg. plantain

racemose inflorescences

numbers show order
of flower opening

simple umbel,
eg. cowslip

umbel

compound umbel,
eg. cow parsley

capitulum

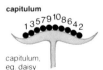

capitulum,
eg. daisy

monandrous (*adj*) describes (1) a flower with only one free stamen; (2) a female animal that mates with only one male.

polyandrous (*adj*) describes (1) a flower with many free stamens; (2) a female animal that mates with several males.

inflorescence (*n*) a group of flowers on the same stalk of a flowering plant. There are two basic types of flower arrangement, racemose (↓) or indefinite inflorescence and cymose (p.70) or definite inflorescence. Inflorescences can be mixed, e.g. a raceme of cymes as in horse-chestnut or compound, e.g. a compound umbel which is an umbel of umbels.

raceme (*n*) an indefinite inflorescence (↑) where the main axis has flowers on stalks, e.g. lupin. In contrast to cymes (p.70) growth in length is from the tip. Flowers open as shown numerically *(see diagrams opposite)* so that the youngest is at the top, or, in flower clusters so that the youngest is at the centre. Panicle (↓), corymb (↓), spike (↓), spadix (↓), catkin (↓), umbel (↓), capitulum (↓) are all types of racemose inflorescence.

panicle (*n*) type of inflorescence (↑), a compound raceme (↑), e.g. oat.

corymb (*n*) type of racemose inflorescence (↑). The lower flower stalks are long so that flowers are all at one level forming a cluster, e.g. candytuft.

spike (*n*) type of racemose inflorescence (↑) in which flowers do not have stalks, e.g. plantain.

spikelet (*n*) a small spike (↑); found in grasses. One or more flowers are enclosed by green scales or **glumes** which often have a stiff process, an **awn**, at the apex.

spadix (*n*) type of inflorescence (↑). A spike (↑) enclosed and protected by a large bract, the **spathe**, e.g. cuckoo pint.

catkin (*n*) type of inflorescence (↑). A spike (↑) of unisexual flowers which often hang down and sway in wind, e.g. hazel.

umbel (*n*) type of racemose inflorescence (↑) in which the axis does not lengthen. Flower stalks arise at the same point so flowers are in a cluster, e.g. cowslip. In compound umbels each flower of a simple umbel is replaced by a small umbel, e.g. cow parsley.

capitulum (*n*) type of racemose inflorescence (↑). The main axis is flattened and bears many sessile flowers, the youngest is in the centre, e.g. daisy.

cyme (*n*) a definite inflorescence (p.69). In contrast to racemes (p.69) the main axis ends in a flower; further development takes place by growth of lateral branches. Each branch of **monochasial** or **simple cymes** bears one other branch which may come off from the same side of the parent stem, e.g. buttercup, or from alternate sides of the parent stem, e.g. iris. **Dichasial** or **double cymes** bear paired branches on either side of the main axis, e.g. stitchwort.

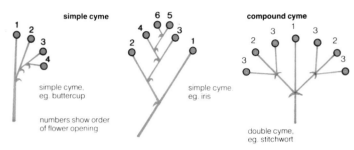

simple cyme

simple cyme,
eg. buttercup

simple cyme.
eg. iris

numbers show order
of flower opening

compound cyme

double cyme,
eg. stitchwort

fruit (*n*) the ripened ovary of a flower. Fruits are formed after fertilization of the ovules. The fertilized ovule develops into a seed within the ripening fruit. Fruits have two scars on them in contrast to seeds (p.74) which have one. Scars on fruits derive from the attachment of the flower stalk and of the style or stigma. Fruits are classified as true or false (p.72) and as simple, aggregate, or multiple (↓). Types of fruit include achene, nut, caryopsis, schizocarp, follicle, legume, silicula, siliqua, capsule (p.72), drupe, berry, pome (p.73). The way in which seeds are dispersed, i.e. by wind, water, animals or the plant's own propulsive mechanism, is related to the type of fruit formed.

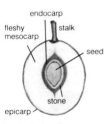

endocarp

fleshy
mesocarp

stalk

seed

stone

epicarp

longitudinal section
of a drupe, plum

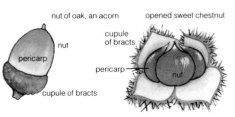

nut of oak, an acorn

nut

pericarp

cupule of bracts

opened sweet chestnut

cupule
of bracts

pericarp

nut

pericarp (*n*) the structure formed from the wall of a plant ovary as the fruit (↑) develops. Can be dry and fibrous (coconut), hard (acorn, sweet-chestnut), or fleshy (blackberry, plum). Fleshy pericarps usually have three distinct layers, the exocarp (↓), mesocarp (↓) and endocarp (↓).

exocarp, epicarp (*n*) outer layer of pericarp (↑) in a fruit. It forms a tough skin round the fruit.

mesocarp (*n*) middle layer of pericarp (↑) in a fruit. It may be fleshy, fibrous or pithy.

endocarp (*n*) inner layer of pericarp (↑) in a fruit. It may be leathery (citrus fruits), stony (drupes), or fleshy and combined with the mesocarp (↑), e.g. berries.

rind (*n*) the outer skin of a succulent fruit, e.g. orange, in which there is a layer of pith under the rind.

flesh, pulp (*n*) the soft part of a succulent fruit, especially the part that is eaten.

stone (*n*) the hard central structure found in some fruits, e.g. drupe of plum. Usually consists of a seed surrounded by an endocarp (↑). *Contrast pip* (↓).

pip (*n*) small hard structure found in succulent fruits, e.g. berries. In contrast to a stone (↑) the seed is not surrounded by an endocarp (↑).

shell (*n*) the hard, outer covering of a fruit, e.g. walnut.

kernel (*n*) (1) the seed within a hard shell (↑). (2) the edible central portion of a nut.

cupule (*n*) a cup-shaped envelope on the fruit of some trees like oak, beech, sweet-chestnut. It is formed from bracts.

burr (*n*) a dry fruit covered in stiff hairs. Hairs are usually hooked. They stick to clothes and fur of animals and aid fruit dispersal.

simple fruit (*n*) type of fruit (↑) formed from a single flower that has one or several fused carpels, e.g. legume, follicle, nut, achene (p.72). *Compare aggregate fruits* (↓), *multiple fruits* (↓).

aggregate fruit (*n*) type of fruit (↑) formed from a single flower with several free carpels, each of which forms a separate ovary. The whole fruit is an aggregate (i.e. a mass or group) of many small, simple fruits (↑), each formed from one ovary, e.g. raspberry, blackberry. *Compare simple fruits* (↑), *multiple fruits* (↓).

multiple fruit, composite fruit (*n*) type of fruit formed from an inflorescence, e.g. pineapple, fig, mulberry. It is a pseudocarp. *Compare simple and aggregate fruits* (↑).

aggregate fruit
an aggregate fruit, blackberry

single
small drupe
or drupel

epicarp

many simple fruits
forming an
aggregate

fruit formed
from fleshy
inflorescence

multiple fruit
a multiple fruit, pineapple

true fruit (*n*) a fruit formed only from the ovary of a flower. *Contrast pseudocarp* (↓).

pseudocarp, false fruit (*n*) a fruit formed from the ovary plus other parts of the flower which develop as a result of fertilization. That of the strawberry and apple includes the receptacle; that of the pineapple and the fig includes the inflorescence.

achene (*n*) a dry, one-seeded, indehiscent (p.75) fruit, e.g. buttercup, clematis. The pericarp is often leathery. In some fruits of this type, e.g. sycamore, the pericarp is extended to form a membranous wing which aids in dispersal. These are called **winged fruits** or **key fruits**.

nut (*n*) a dry, one-seeded, indehiscent (p.75) fruit. The pericarp forms a hard, woody wall or shell, e.g. hazel nut, cashew nut, acorn, sweet-chestnut.

caryopsis (*n*) a dry, one-seeded, indehiscent (p.75) fruit in which the pericarp is fused with the testa of the seed. Found in grasses, e.g. maize.

schizocarp (*n*) a dry fruit formed from several united carpels. When ripe it splits into segments which are usually one-seeded, e.g. geranium, hollyhock.

follicle (*n*) dry, dehiscent (p.75) fruit formed from a single carpel. It splits along one side only to liberate seeds, e.g. delphinium. *Contrast legume* (↓).

legume, pod (*n*) dry, dehiscent (p.75) fruit formed from a single carpel. It contains several seeds. A legume splits along two sides to liberate seeds, e.g. pea, bean. *Contrast follicle* (↑).

silicula (*n*) dry, dehiscent (p.75) fruit formed from two fused carpels. It is short and fat. *Contrast siliqua* (↓). A central false septum between the carpels divides the fruit into two compartments. When ripe the carpel walls split off the false septum from below to liberate the seeds, e.g. honesty, shepherd's purse.

siliqua (*n*) a dry, dehiscent (p.75) fruit formed from two fused carpels. It is long and cylindrical. *Contrast silicula* (↑). A central false septum between the carpels divides the fruit into two compartments. When ripe the carpel walls split off the false septum from below to liberate the seeds, e.g. stock.

capsule (*n*) a dry, dehiscent (p.75) fruit formed from two or more fused carpels. It contains many seeds and opens to liberate them in several ways, e.g. by splits (carnation, nigella), pores (poppy, antirrhinum) or lids (scarlet pimpernel).

achene

achene of clematis

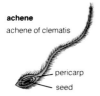

pericarp

seed

winged fruit

winged fruits of sycamore

pericarp extended to form membranous wing

pericarp within which lies one seed

follicles

follicles of delphinium

line where follicle has split to release seeds

legume pod

pericarp segments twist on drying

seed

pod split along two sides

legume of garden pea

seed beneath
pericarp and
attached to
septum

false septum
to which seeds
are attached

silicula   pericarp
split from
below

silicula of honesty

false septum
to which seeds
are attached

pericarp
segments
twist on
drying

pericarp split
from below

siliqua

siliqua of stock

pores to allow seeds
to be liberated
when wind blows

capsule

capsule of poppy

drupe, stone fruit (n) a succulent fruit formed from one
carpel. A single seed is enclosed by a pericarp
consisting of a thin, skin-like epicarp, a fleshy
mesocarp and a hard, stony endocarp, e.g. plum,
mango, cherry. The coconut is a drupe with a fibrous
mesocarp. *Contrast berry* (↓).

drupel (n) a small drupe (↑); found in aggregate fruits
(p.71).

berry (n) a succulent fruit formed from a single flower
with several fused carpels. It has a thin epicarp. In
contrast to a drupe (↑) it does not have a stony
endocarp. The endocarp is fleshy and combined with
the mesocarp. It may have one seed (date) or several
(tomato, banana, gooseberry, papaya). Each seed is
inside a tough testa (p.74).

pome (n) a pseudocarp (↑). Most of the fruit is developed
from the receptacle of the flower which becomes
swollen and fleshy to enclose the true fruit, e.g. apple.
The edible, fleshy part of the apple is formed from the
receptacle. The apple core is the true fruit formed from
the ovary; it contains several seeds.

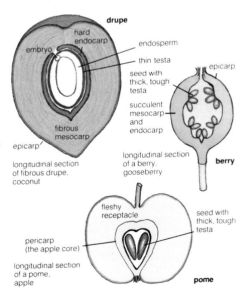

drupe

hard
endocarp

embryo

fibrous
mesocarp

epicarp

longitudinal section
of fibrous drupe,
coconut

endosperm

thin testa

seed with
thick, tough
testa

succulent
mesocarp
and
endocarp

epicarp

longitudinal section
of a berry,
gooseberry

berry

fleshy
receptacle

pericarp
(the apple core)

longitudinal section
of a pome,
apple

seed with
thick, tough
testa

pome

seed (*n*) a structure formed from a fertilized ovule. It contains an embryo ready for germination, surrounded by a protective seed coat or testa (↓). In endospermic seeds, e.g. castor oil, the embryo is surrounded by a food supply, i.e. the endosperm (↓) tissue. Non-endospermic seeds, e.g. peas, beans, do not have endosperm. Unlike fruits (p.70), which have two scars, seeds only have one, the hilum (↓) marking the point of attachment of the ovule stalk. Most seeds have a micropyle (↓) and some have additional structures, e.g. a caruncle (↓).

endosperm (*n*) a tissue which surrounds and nourishes the embryo of seed plants. In non-endospermic seeds (e.g. peas, beans) the endosperm is absorbed into the cotyledons of the embryo by the time the seed is fully developed. Part of the endosperm remains in endospermic seeds (e.g. castor oil) to provide a store of food for the seedling.

testa, seed coat (*n*) the external, protective covering of a seed (↑) which is usually hard and dry. It is formed from the integument(s) of the ovule.

hilum (*n*) the scar left on a seed (↑). It marks the former point of attachment of the seed to the funicle (ovule stalk).

micropyle (*n*) a small opening in the seed coat, water enters through it when the seed germinates. It is often the site of entry of the pollen tube to the ovule so fertilization can occur.

caruncle (*n*) an outgrowth from the testa (↑) of a few flowering plants, e.g. castor oil; it hides the micropyle (↑).

aleurone layer (*n*) the outer layer of the endosperm (↑) of cereal seeds. The cells contain aleurone grains (↓).

aleurone grain (*n*) granules of protein found in plants, e.g. in the aleurone layer (↑) of cereal seeds. It stores mainly proteins and the enzymes to digest food stores on germination.

gluten (*n*) a reserve protein found in cereals, particularly wheat.

grain (*n*) (1) a small, hard particle of matter, e.g. the aleurone grain of seeds (↑); (2) a single small, hard seed; (3) collective name for the seeds of a cereal, e.g. maize, wheat.

husk (*n*) term used for the dry, thin covering of some seeds. Husks are removed from the grain (↑) of cereals like maize and wheat before it is ground into flour.

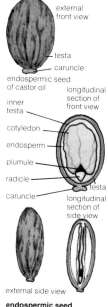

external front view

testa

caruncle

endospermic seed of castor oil

longitudinal section of front view

inner testa

cotyledon

endosperm

plumule

radicle

testa

caruncle

longitudinal section of side view

external side view

endospermic seed

external front view

external side view

position of radicle

micropyle

endospermic seed of broad bean

hilum

cotyledon

plumule

radicle

front view without testa

groove for plumule

plumule

radicle

side view, testa removed and cotyledons separated

dispersal of fruits and seeds (*n*) the scattering of fruits and seeds over a wide area to increase distribution. This increases the chance of finding nourishment and suitable conditions for the species to survive. Dispersal may be by vectors, e.g. wind (↓), water (↓), animals (↓) or by the plant's own propulsive mechanism (↓).

wind dispersal (*n*) type of dispersal (↑) associated with very light seeds and fruits, e.g. willow herb, dandelion. Wind also disperses the small spores of fern and fungi. Heavier fruits and seeds may have wings to help in wind dispersal, e.g. sycamore, conifers, elm.

censer mechanism (*n*) a method of seed dispersal (↑) using wind as the vector. Seeds are thrown from a capsule, through pores, when the wind blows, e.g. delphinium, campion, poppy.

parachute (*n*) a structure on a fruit or seed that reduces the rate at which it descends. It helps dispersal by wind, e.g. dandelion, willow herb.

water dispersal (*n*) type of dispersal (↑) associated with seeds and fruits that have fibrous layers, e.g. coconut, or spongy structures that enable them to float, e.g. water lily. This type of dispersal is not common.

animal dispersal (*n*) type of dispersal of fruits and seeds involving animals. Fruit or seeds, e.g. cherry, apple, orange may be eaten by them; seeds are then passed out with the faeces. Some fruits and seeds have hooks which attach them to the skin or fur of an animal or clothes of man, e.g. agrimony, burdock. Those like blackberry that are eaten by birds may stick to the beak and thus be transported. Rodents, e.g. squirrels, disperse nuts when transporting them to their food stores.

propulsive mechanism (*n*) method of seed dispersal (↑) that does not involve a vector. When the fruit is mature structures within the plant enable seeds to be forcibly expelled some distance from the plant, e.g. unequal drying of the pericarp in the legume of peas and the siliquas of stock and turgidity of the pericarp in balsam.

dehiscent (*adj*) describes plant structures, particularly dry, mature fruits which open to liberate seeds, e.g. peas. The pattern of dehiscence is characteristic for a particular fruit, e.g. pores in poppy; splitting along septa in honesty. *Contrast indehiscent* (↓).

indehiscent (*adj*) describes a dry fruit that does not open spontaneously to liberate seeds when ripe, e.g. hazel, strawberry, sycamore. *Contrast dehiscent* (↑).

parachutes

fruit parachute of dandelion

seed parachute of willow herb

mature (*adj*) describes something which is fully developed or ripe (↓), e.g. wine and cheese that are ready for consumption; fruit with seeds ready for dispersal; a human that is fully developed sexually. *Compare immature* (↓).

immature (*adj*) describes something which is not fully developed, is not mature and is not ripe, e.g. a fruit with seeds that are not fully formed, a young child before puberty. *Compare mature* (↑).

ripe (*adj*) describes something which is fully developed and mature (↑), e.g. a ripe fruit that has seeds ready for dispersal.

germination (*n*) the growth of an embryo (↓) of a viable seed or spore in plants. It often follows a period of dormancy and begins in response to favourable conditions of warmth, sufficient oxygen, water and sometimes the presence of light.

imbibition (*n*) the absorption of water; e.g. seeds imbibe water just before germination (↑). The increase in volume causes the testa to split.

epigeal (*adj*) describes seed germination (↑) in plants in which cotyledons (↓) appear above ground, e.g. castor oil, sunflower, pine, onion. *Compare hypogeal* (↓).

hypogeal (*adj*) describes seed germination (↑) in plants in which the cotyledons (↓) remain below ground, e.g. maize, beans, peas. *Compare epigeal* (↑).

embryo (*n*) in plants, the structure within the seed that will develop into the new plant. It derives from the fertilized egg cell. An embryo consists of a plumule (↓), radicle (p.53) and one or more cotyledons (↓).

cotyledons, seed leaf (*n*) a leaf-like part of the embryo (↑) of seed plants. Its structure is much simpler than that of the foliage leaves formed later. When in the seed a cotyledon does not contain chlorophyll, therefore is not green. **Monocotyledons** have one, **dicotyledons** have two, and gymnosperms have several cotyledons in each seed. Cotyledons in some seeds, e.g. pea, bean, are storage organs from which the seedling draws food. In others, e.g. grasses, maize, caster oil, food is stored in the endosperm (p.74), absorbed by the cotyledons and passed on to the seedling. Cotyledons of epigeal (↑) plants appear above ground, develop chlorophyll and carry out photosynthesis.

scutellum (*n*) (1) part of the embryo of grasses. (2) the large cotyledon (↑) in maize seeds.

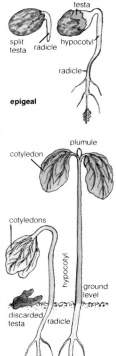

ground level

inner testa

split testa   radicle   hypocotyl

radicle

epigeal

plumule

cotyledon

cotyledons

hypocotyl

ground level

discarded testa / radicle

secondary root

epigeal germination of castor oil, a dicotyledon

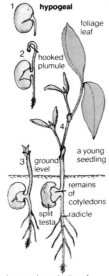

hypogeal germination of
broad bean, a dicotyledon

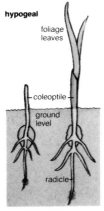

hypogeal germination
of maize, a monocotyledon

plumule (*n*) terminal bud of the embryo (↑) in a seed
plant. The stem of the seedling develops from it.

coleoptile (*n*) a protective sheath surrounding the
plumule (↑) of the embryo in grasses. It forms the first
leaf of the aerial shoot.

seedling (*n*) a young plant just formed from a
germinating (↑) seed.

hypocotyl (*n*) the part of the stem of a seedling that lies
below the point of attachment of the stalk of the
cotyledon (↑) and above the primary root. The structure
of the vascular tissue changes in the hypocotyl from
that of a root to that of a stem. *Compare epicotyl* (↓).

epicotyl (*n*) the part of the seedling stem lying above the
point of attachment of the cotyledon (↑) stalk. *Compare
hypocotyl* (↑).

vegetative reproduction, – propagation (*n*) asexual
reproduction in plants involving development of part of
the plant body, i.e. root, stem or leaf, into a new and
ultimately separate plant. Organs of vegetative
reproduction include rhizomes, tubers, corms, bulbs,
bulbils, stolons, runners, offsets, suckers (p.78).
Advantages of vegetative propagation over production
of a new plant from seeds (sexual reproduction) are:
it rapidly colonizes an area; fruits and seeds are obtained more rapidly; propagated plants are stronger; offspring are genetically the same as the parent
(important in plant breeding to maintain selected
characteristics). This is the only method of propagating
seedless plants.

perennation (*n*) any method of overwintering in plants.
Annuals perennate by means of seeds; biennials by
storing food in underground storage organs, e.g. tap
roots; herbaceous perennials by underground storage
organs, e.g. bulbs, corms, rhizomes and tubers (p.78)
which can also be organs of vegetative reproduction (↑);
woody perennials overwinter by using food stored in
woody tissue. Deciduous trees also shed their leaves;
in evergreens metabolic activity is reduced.

dormancy (*n*) a resting condition of plants in which
growth stops and metabolism is minimal. Enables
organisms to survive adverse conditions. Dormancy
involves a reproductive body only, e.g. spores of seed
plants, or the whole organism, e.g. bulbs, corms, tubers
of higher plants. *Compare hibernation (p.185) in
animals.*

rhizome (*n*) a horizontal underground plant stem that bears buds, leaves, adventitious roots, e.g. iris. It is a means of vegetative reproduction (p.77) and in some plants a perennating organ (p.77).

tuber (*n*) an organ of perennation (p.77) and vegetative reproduction (p.77) in plants. A stem tuber is the swollen end of an underground stem, e.g. the potato. It stores food and bears buds in the axils of scale leaves. Each bud can grow into a new plant. A root tuber is a swollen root that acts as a food store, e.g. dahlia.

corm (*n*) an underground, swollen, rounded base of a plant stem (e.g. crocus, gladiolus) that stores food. It is covered with scale-like remains of leaves from the previous year's growth; they have buds in their axils. If one bud only grows, the corm is a perennating organ; if more grow it acts also as an organ of vegetative reproduction. *Compare bulb* (↓).

bulb (*n*) a plant organ of perennation and normally also vegetative propagation, e.g. onion, tulip. Bulbs are modified shoots consisting of thick, very much shortened stems surrounded by fleshy scale leaves or thickened leaf bases. They have buds in their axils. If one bud grows, the bulb is a perennating organ; if more grow it is also an organ of vegetative reproduction. *Compare corm* (↑).

bulbil (*n*) a very small, fleshy bulb-like organ of vegetative reproduction (p.77) arising from aerial structures of some plants, e.g. certain species of lily. It drops off and forms a new plant.

stolon (*n*) in plants an organ of vegetative reproduction (p.77). It is a stem growing horizontally on the surface of the soil that takes root at nodes and forms new plants. Long stolons that root at the tip are called runners (↓), e.g. strawberry; short stolons are offsets, e.g. houseleek.

runner (*n*) a short stolon (↑) that grows rapidly in length, roots at the tip and forms a new plant. The runner then decays, e.g. strawberry. Runners differ from **creepers** (p.54) in that the main stem bears scale leaves in the axils of which foliage leaves arise. Creepers, e.g. ground ivy, have foliage leaves arising directly from nodes.

sucker (*n*) an organ of vegetative propagation (p.77) in plants. It develops from an underground stem which becomes an aerial shoot. It often emerges some distance from the main stem, roots develop and a new plant is formed, e.g. mint.

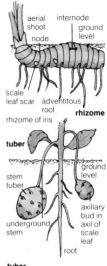

rhizome of iris
rhizome

tuber
root tubers of dahlia
tuber

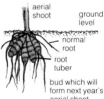

remains of last year's corm
corm
crocus corm with half of scale leaves removed

longitudinal section of tulip bulb
bulb

offset   offset of houseleek

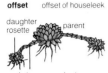

daughter rosette   parent

scale leaves on short stolon or runner

creeper
creeper of ground ivy

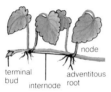

node

terminal bud   adventitous root

internode

sucker
sucker of mint

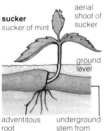

aerial shoot of sucker

ground level

adventitous root   underground stem from parent plant

stock   **budding**

scion

T-shaped cut

grafting

horticultural propagation, artificial propagation (n) vegetative propagation in plants brought about artificially, e.g. by cuttings (↓), layering (↓), budding (↓), grafting (↓).

cutting (n) a piece of branch, stem, root or a leaf which when planted grows into a new individual. It is a means of artificial propagation (↑).

layering (n) a method of artificial propagation (↑). A plant branch is bent so either the tip or part of it can be covered with soil. After adventitious roots have grown at the node, the stem can be cut free from its parent.

grafting (n) (1) a method of artificial propagation (↑) in plants; particularly used in propagation of roses and fruit trees. A small part of one plant is transplanted on to another. The plant from which the part is transferred is the scion (↓); the one receiving it is the stock (↓). A shoot of the scion is normally grafted on to the lower part of the shoot of the stock. Thus the root system is formed from the stock and the shoot system from the scion. For a successful graft the cambium of both the scion and the stock must be in direct contact. Only closely related plants can be grafted. When only buds of scion are grafted, the process is called budding (↓). (2) grafting is also possible between animals. A small part, usually skin, kidney, heart, of one individual, the donor, is grafted on to another animal, the host or recipient.

graft (n) small part of an animal or plant that is grafted (↑) on to either another individual or to a different position on the same individual.

scion (n) the part of a plant grafted (↑) on to the stock (↓) of another. It forms the shoot of the new plant.

stock (n) part of a plant, usually the root and the lower part of the stem on to which a scion (↑) is grafted. It forms the root of the new plant.

budding (n) method of artificial propagation (↑) in plants. A form of grafting (↑) in which the grafted part is a bud. A T-shaped cut is made in the stock (↑) into which the bud of the scion (↑) is placed. A new shoot forms from the bud. Budding is used particularly in rose, peach, plum cultivation.

scions   completed graft

stocks

body (*n*) the whole of an animal, usually excluding the head and sometimes the other appendages.

trunk (*n*) (1) the body (↑) of an animal excluding head and limbs. (2) the main part of a structure that bears branches, e.g. a nerve trunk before it branches. (3) the proboscis of an elephant.

torso (*n*) the trunk (↑) of a human, i.e. the body (↑) excluding head and limbs.

limb (*n*) a vertebrate appendage, e.g. arms and legs of man, fore and hind limbs of animals, wings of birds.

fore limb, fore leg (*n*) the front leg of an animal.

hind limb, hind leg (*n*) the back leg of an animal.

shoulder (*n*) region of body of a vertebrate to which a fore limb (↑) or arm (↓) is attached.

arm (*n*) fore limb (↑), particularly of a human.

forearm (*n*) the part of the arm (↑) between the elbow (↓) and wrist (↓).

elbow (*n*) joint between the forearm (↑) and upper arm.

wrist (*n*) joint between hand (↓) and forearm (↑).

hand (*n*) in man the end of the arm below the wrist (↑). It terminates in fingers (↓).

palm (*n*) the inner surface of the hand (↑) between wrist (↑) and fingers (↓).

finger (*n*) terminal part of the hand (↑), a digit (↓). There are four fingers and one thumb (↓); they have nails (↓).

thumb (*n*) a short, thick digit (↓) on the radial side of the human hand (↑).

digit (*n*) one of the terminal divisions of a pentadactyl limb, e.g. fingers and toes in man.

nail (*n*) a horny plate at the end of the digits (↑) of some mammals. Formed by the epidermis. Distinguished from claws by being flattened.

hip (*n*) joint at which a hind limb (↑) or hind leg (↑) is attached to the body.

thigh (*n*) thick, fleshy, upper part of a leg (↓) between the hips (↑) and the knee (↓).

leg (*n*) (1) hind limb of human; (2) limb of vertebrate, e.g. leg of horse; (3) locomotory appendages of animals generally, e.g. legs of insect.

shin (*n*) the front part of a leg (↑) below the knee (↓) of some animals including man.

calf (*n*) the thick, fleshy part at the back of a leg (↑) below the knee of some animals including man.

knee (*n*) joint between the thigh (↑) and lower part of the leg (↑) of some vertebrates, including man.

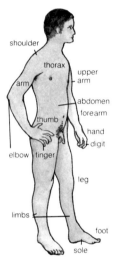

anatomy of human male

ankle (*n*) joint between foot (↓) and leg (↑).
foot (*pl.feet*) (*n*) the end of the leg (↑) below the ankle (↑).
It terminates in toes (↓). The part of a body on which an
animal stands or walks.
heel (*n*) hind part of foot (↑) below the ankle (↑).
sole (*n*) underside of foot (↑) between heel (↑) and toes.
toe (*n*) terminal part of the foot (↑), a digit (↑).
head (*n*) end or foremost part of an animal body. The
brain is located in it.
scalp (*n*) top, hairy part of the head (↑); covers the skull.
cheek (*n*) (1) fleshy side wall of the mouth. (2) the side of
the face below the eye.
neck (*n*) part of body connecting head (↑) and trunk (↑).
chest (*n*) (1) upper, front part of human body between
neck (↑) and abdomen (↓), i.e. the breast. (2) thorax in
humans.
breast (*n*) (1) upper, front part of an animal body
between neck and abdomen (↓). (2) in women one of
the two mammary glands on the chest (↑).
thorax (*n*) the part of the body between head and
abdomen (↓), especially in arthropods and mammals. In
mammals it is separated from the abdomen by the
diaphragm and contains the heart and lungs. In insects
it carries legs and wings *(see p.39)*. **thoracic** (*adj*).
abdomen (*n*) (1) posterior part of the body of arthropods;
(2) the body cavity in vertebrates that contains the
intestines, liver, kidneys. In mammals it is separated
from the thorax by the diaphragm. **abdominal** (*adj*).
midriff (*n*) the diaphragm of a mammal.
navel, umbilicus (*n*) the depression in the centre of the
abdomen of a mammal; marks former point of
attachment of the umbilical cord.
body cavity (*n*) internal cavity of animals in which lie the
viscera (↓).
viscera (*n*) the organs in the body cavity (↑) of an animal;
especially the intestines and associated organs.
coelom (*n*) main body cavity where body is made up of
three layers. In higher animals, e.g. man, divided into
separate cavities, i.e. peritoneal, pleural, pericardial
cavities.
peritoneal cavity (*n*) body cavity (↑) of the abdomen in
animals. Contains liver, spleen, kidney and most of gut.
pleural cavity (*n*) body cavity (↑) of the thorax in mammals.
There are two, each lined by a pleural membrane (a **pleura**)
and containing one of the lungs.

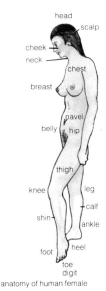

head
scalp
cheek
neck
chest
breast
navel
belly
hip
thigh
knee
leg
calf
shin
ankle
foot
heel
toe
digit
anatomy of human female

organ (*n*) part of an animal or plant that is composed of various tissues. Forms a structural and functional unit, e.g. heart, lung, leaf, root.

system (*n*) set of organs in a plant or animal concerned with a particular function, e.g. circulatory system.

bilaterally symmetrical (*adj*) an object which can be halved in one plane only forming two mirror images, e.g. all vertebrates. It is common in animals but rare in plants; in flowers this state is called **zygomorphy** (p.62). *Compare radially symmetrical (↓).*

radially symmetrical (*adj*) an object, e.g. a cylinder, which can be halved along many planes through its axis forming two halves which are mirror images. In flowers this is called **actinomorphy** (p.62). *Compare bilaterally symmetrical (↑).*

axis (*n*) line through a structure dividing it into two halves.

anterior (*adj*) describes (1) the head or front of an animal. In man it is the ventral surface (↓); (2) the part of a lateral bud or flower furthest away from the main axis. Opposite of posterior (↓).

ventral (*adj*) lying on or near the lower surface, especially of animals. In erect animals, e.g. man, its equivalent is the anterior (↑) or front. Opposite of dorsal (↓).

superior (*adj*) used in animals with erect posture, e.g. man, instead of anterior (↑) to describe parts near the head end. Opposite of inferior (↓).

posterior (*adj*) describes (1) parts of an animal furthest away from the front or head. In man it is the back, dorsal surface (↓); (2) the part of a lateral bud or flower nearest the main axis. Opposite of anterior (↑).

dorsal (*adj*) lying on or near the upper surface, especially of animals. In vertebrates the back bone is in this region. In erect animals, e.g. man, it is the back or posterior (↑) surface. Opposite of ventral (↑).

inferior (*adj*) used in animals with erect posture, e.g. man, instead of posterior (↑) to describe parts furthest from the head. Opposite of superior (↑).

proximal (*adj*) describes the part of a structure in plants or animals nearest to the attached end, e.g. the humerus is the proximal part of the arm. *Compare distal (↓).*

distal (*adj*) describes the part of a structure in plants or animals furthest away from the attached end, e.g. the hand is the distal part of the arm. *Compare proximal (↑).*

lateral (*adj*) describes the part of a plant or animal lying at the side.

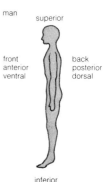

anatomical terms of position

man

superior

front
anterior
ventral

back
posterior
dorsal

inferior

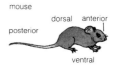

mouse

dorsal   anterior

posterior

ventral

proximal

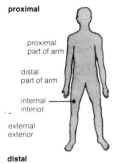

proximal
part of arm

distal
part of arm

internal
interior

external
exterior

distal

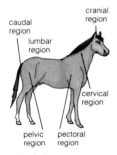

cranial
region

caudal
region

lumbar
region

cervical
region

pelvic   pectoral
region   region

anatomical terms
of position

medulla
cortex

section of kidney

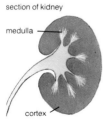

medulla

cortex

peripheral (*adj*) describes a part of a plant or animal lying near the outer surface called the **perifery** (*n*).

terminal (*adj*) describes a part lying at the end of a structure, e.g. fingers are the terminal part of the hand.

appendicular (*adj*) concerning an appendage, e.g. arm.

external (*adj*) describes something outside the structure being considered. *Compare internal* (↓).

internal (*adj*) describes something inside the structure being considered. *Compare external* (↑).

cranial (*adj*) concerned with the head of an animal.

labial (*adj*) concerned with the lips of a structure.

oral (*adj*) concerned with the mouth of a structure.

orifice (*n*) a small, mouth-like opening.

cervical (*adj*) concerned with the neck of an animal or organ, e.g. part joining head and trunk.

lumbar (*adj*) concerned with the region of an animal below the chest and above the pelvis.

pectoral (*adj*) concerned with the chest of an animal.

pelvic (*adj*) concerned with the hip of an animal.

caudal (*adj*) concerned with the tail of an animal.

flesh (*n*) the soft, living substance of a plant or an animal, e.g. muscle (meat) of animals.

adipose tissue (*n*) fatty tissue. Occurs especially in mammals, mainly under the skin, in the mesenteries and around the kidneys. It stores fat which provides insulation and can be converted into energy.

foramen (*n*) an opening in an animal body, e.g. the opening to a canal in bone; the opening in a vertebrate skull through which the spinal cord passes.

sphincter (*n*) a ring of muscle in the wall of a tube or at the opening of a hollow organ. It contracts to narrow or close the orifice; when relaxed the orifice (↑) is open.

medulla (*n*) central, inner part of an animal tissue or organ, particularly of the adrenal glands and kidney.

cortex (*n*) outer layer of an animal organ, particularly of the cerebrum, adrenal glands and kidney.

gland (*n*) in animals an organ that makes specific chemical substances for secretion. Exocrine glands (p.84) secrete their products through a duct either on to the outer surface of the body (e.g. sweat glands) or internally (e.g. digestive glands on to the inner surface of the gut). Endocrine glands (p.180) secrete directly into blood.

duct (*n*) a tube in animals and plants to carry fluid from one system to another, e.g. bile is carried from the liver to the alimentary canal in the bile duct.

channel (*n*) a passage for movement of substances, usually liquids.

exocrine gland (*n*) animal gland (p.83) with a duct to secrete its products either on to the outer surface of the body or internally.

follicle (*n*) in animals a small cavity or gland.

sinus (*n*) (1) a small cavity in a tissue, e.g. nasal sinus, lymphatic sinus; (2) a dilated tube, e.g. a blood sinus.

sac (*n*) a bag-like container or pouch, e.g. pollen sacs of flowers contain pollen.

structure (*n*) arrangement of parts that make up a cell, tissue or organ.

function (*n*) the way an organism or part of it acts to carry out the vital processes in which it is involved.

position (*n*) the place occupied relative to other organisms or parts of organisms (↓).

organism (*n*) the whole plant or animal. It is composed of organs which work together to support life.

type (*n*) an individual with essential (↓) properties and general (↓) characteristics of the group to which it belongs.

essential, vital (*adj*) describes a part or process which is necessary to support life.

general (*adj*) describes characteristics found in all the members of a group. *Contrast particular* (↓).

particular (*adj*) describes characteristics found only in an individual of the group. *Contrast general* (↑).

normal (*adj*) describes something ordinary or average, i.e. it is not different in any way from others of the group. Also a process which is working correctly.

abnormal (*adj*) describes something which is not normal.

analogous (*adj*) describes organs or structures which although having similar functions do not have similar structures and evolutionary origins, e.g. beak of bird and mouth of animal are analogous, both being involved in feeding but they are structurally different. *Contrast homologous* (↓).

homologous (*adj*) describes organs or structures which although having similar structure and evolutionary origin do not necessarily have the same functions, e.g. bird wing and horse's leg. *Contrast analogous* (↑).

layer (*n*) a stratum (↓) or region of material covering a structure, e.g. epidermal layer of skin.

stratum (*n*) (1) a layer (↑) of cells in living tissue, e.g. stratum corneum in skin; (2) a bed of sedimentary rock; (3) region of vegetation of similar height.

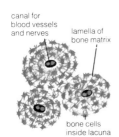

canal for
blood vessels
and nerves   lamella of
              bone matrix

                    bone cells
                    inside lacuna

bone

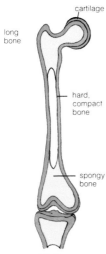

cartilage

long
bone

                    hard,
                    compact
                    bone

                    spongy
                    bone

bones
articulating
with one
another

bundle (*n*) a number of things loosely bound together, e.g. a strand of nerve fibres in animals or a bundle of vascular tissue in plants.

filament (*n*) a very thin, thread-like object, e.g. a fibre, stalk of a stamen, chain of cells, hypha of fungus.

attach (*v*) to connect, join, or fasten a whole or part of a structure to another structure.

attachment (*n*) the joining of a tissue or organ in plants and animals to other tissues or organs, e.g. the connective tissue by which striped muscle is joined to bone.

origin (*n*) (1) point at which a muscle is attached to bone; (2) start or first existence from which something has developed.

bone (*n*) a connective tissue forming the skeleton of vertebrates. Consists of cells embedded in a matrix of bone salts and collagen fibres. The bone salts, mostly calcium carbonate and calcium phosphate, make up about 60% of its mass and give it hardness. Collagen fibres give it tensile strength. Bone cells are connected by fine channels running through the matrix; larger channels contain blood vessels and nerves. Bone supports and protects body organs and acts as a store of calcium and phosphate ions. There are several bone types, compact bone (↓), spongy bone (↓) and cartilage bone (↓).

compact bone (*n*) type of bone (↑). It forms the hard, outer layer of the long bones of adult humans; it has fewer cavities than spongy bone (↓).

spongy bone (*n*) type of bone (↑). It is found in bones of young vertebrates and internal to the compact bone (↑) in the long bones of adult humans. It is the first stage in the ossification (p.93) of cartilage and cavities are left as the cartilage dies away.

cartilage bone (*n*) bone (↑) type formed by ossification (p.93) of cartilage (p.92). It makes up the bones of most young vertebrates. All limb bones, girdle bones, vertebral column, and the bones of the base of the skull are cartilage bones.

bone fracture (*n*) break or crack in a bone. There are several types. **Simple** in which the bone is broken and there is only slight injury to surrounding tissue. **Compound** involving an external wound leading down to the fracture; broken bone ends may protrude through the skin. **Complicated** involves internal injury to tissues, blood vessels or organs around the fracture. **Impacted** where broken bone ends are driven into one another.

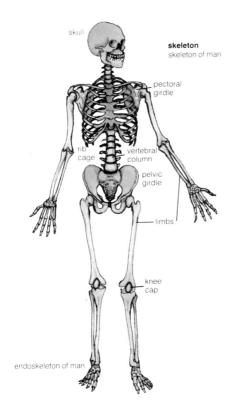

skull

skeleton
skeleton of man

pectoral
girdle

rib
cage

vertebral
column

pelvic
girdle

limbs

knee
cap

endoskeleton of man

skeleton (*n*) the hard parts that form the framework of an
animal body. It provides support and protects internal
organs, gives shape to the body and provides
anchorage points for muscles and levers for movement.
There are two basic types, exoskeletons (↓) and
endoskeletons (↓).

articulation (*n*) the ability of two parts of the skeleton of
an animal to move relative to each other. Usually it
involves a joint.

articulated (*adj*) describes a structure which is attached
to another structure by a movable joint, e.g. the leg
bones are articulated at the knee.

exoskeleton (*n*) a skeleton (↑) that either lies outside the body tissues of an animal, e.g shell of molluscs, cuticle of insects, or is situated in the skin, e.g. bony plates of tortoise, armadillo, scaly fish. It protects and supports internal organs and may provide attachment for muscles. *Compare endoskeleton* (↓). ·

endoskeleton (*n*) a skeleton (↑) that lies inside the body tissues of an animal, e.g. bony skeleton of many vertebrates. It protects and supports internal organs, gives shape to the body and provides anchorage points for muscles. Some animals, e.g. turtle, may have both an endoskeleton and an exoskeleton (↑).

axial skeleton (*n*) part of the endoskeleton (↑). It consists of the skull and vertebral column to which is attached the rib cage.

appendicular skeleton (*n*) part of the endoskeleton (↑). It consists of four limbs attached to the bony pectoral and pelvic girdles.

pentadactyl limb (*n*) the type of limb found in amphibia, reptiles, birds, mammals, i.e. all vertebrates except fish. It evolved as an adaptation to terrestrial life. Although modified in some species all have the same basic pattern of: (a) an upper part with one long bone; (b) a lower part with two long bones; (c) a part that ends in five digits. In the fore limb these are: (a) the humerus; (b) radius and ulna; (c) metacarpals and phalanges. In the hind limb they are: (a) the femur; (b) tibia and fibula; (c) metatarsals and phalanges.

pentadactyl limb

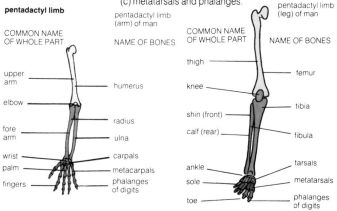

pentadactyl limb (arm) of man

COMMON NAME OF WHOLE PART

COMMON NAME OF WHOLE PART	NAME OF BONES
upper arm	humerus
elbow	
	radius
fore arm	ulna
wrist	carpals
palm	metacarpals
fingers	phalanges of digits

pentadactyl limb (leg) of man

COMMON NAME OF WHOLE PART	NAME OF BONES
thigh	femur
knee	
	tibia
shin (front)	
calf (rear)	fibula
ankle	tarsals
sole	metatarsals
toe	phalanges of digits

vertebral column, spinal column, backbone, spine (*n*) all terms for a flexible, jointed column made of bones called vertebrae (↓). It lies near the dorsal surface of a vertebrate animal (e.g. the back or posterior of man) and runs from skull to tail. It provides support for the body; protects and encloses the spinal cord (p.161).

vertebra (*n*) (*pl. vertebrae*) a bony segment of the vertebral column (↑). Vertebrae form a flexible, jointed column running from skull to tail near the dorsal surface of a vertebrate animal. Each vertebra articulates with those on either side; only restricted movement is possible between any two vertebrae. Discs of cartilage between vertebrae absorb shock. A vertebra consists of a mass called the centrum (↓) with a neural arch above forming a spinal canal which encloses and protects the spinal cord. A transverse process projects on either side of the neural arch; between it and the centrum is a hole to provide exits for spinal nerves. In some vertebrae there is a long process from the neural arch called the neural spine (↓). In fish all the vertebrae are similar. In most other vertebrates differences occur in vertebrae from different parts of the spinal cord, see axis (↓), atlas (↓), cervical (↓), thoracic (↓), lumbar (↓), sacral (↓) and caudal (↓) vertebrae. In birds many are fused to give rigidity for flight.

centrum (*n*) thick mass of bone in a vertebra (↑). Each is firmly but flexibly attached to adjacent centra forming the main support of the backbone. A neural arch (↓) arises from it.

neural arch (*n*) a ring of bone arising from the centrum (↑) of a vertebra so that it forms the spinal canal (↓). Transverse processes and in some vertebrae a neural spine (↓) arise from it.

neural spine (*n*) long bony projection from the neural arch (↑) of a vertebra (↑). Serves as a muscle attachment.

transverse process (*n*) projections on either side of the neural arch (↑) of a vertebra (↑). In thoracic vertebrae ribs articulate with them.

zygapophysis, anterior and posterior (*n*) small flanges of bone at the front and back of a vertebra by which two adjacent vertebrae lock into one unit.

spinal canal, neural canal (*n*) canal formed by the neural arch (↑) of a vertebra (↑) where it arises from the centrum (↑). It encloses and protects the spinal cord.

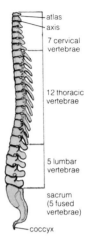

atlas
axis
7 cervical vertebrae

12 thoracic vertebrae

5 lumbar vertebrae

sacrum (5 fused vertebrae)

coccyx

vertebral column
vertebral column of man

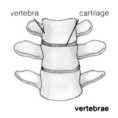

vertebra   cartilage

vertebrae

neural spine   zygapophysis

transverse process

neural arch

neural canal

vertebra

vertebraterial canal (*n*) canal in cervical vertebrae (↓) for passage of the vertebral artery.

atlas (*n*) first vertebra (↑) of the backbone in amphibians, reptiles, birds, mammals. It is modified to allow free movement of the head. Skull-atlas joint allows movement up and down; atlas-axis joint allows head to rotate.

axis (*n*) second vertebra (↑) of backbone of reptiles, birds, mammals. The odontoid process (↓) on it projects into the atlas (↑) allowing the head to rotate.

odontoid process, – peg (*n*) projection on the axis (↑) that projects into the atlas (↑) vertebra to allow rotation of the head.

cervical vertebrae (*n.pl.*) bones in the neck region of the backbone (↑). They have vertebraterial canals (↑).

thoracic vertebrae (*n.pl.*) bones in the chest region of the backbone (↑). They articulate with the ribs.

lumbar vertebrae (*n.pl.*) bones in the abdominal region of the backbone (↑). They have large, broad, transverse processes (↑).

sacral vertebrae (*n.pl.*) bones in the lower region of the backbone (↑). They articulate with the pelvic girdle. In some animals, e.g. man, they are fused together forming the **sacrum**.

caudal vertebrae (*n.pl.*) bones in the tail region of the backbone (↑).

coccyx (*n*) fused caudal vertebrae of tailless primates, e.g. man.

skull (*n*) most anterior part of the skeleton consisting of cranium (↓) and facial skeleton.

cranium (*n*) part of the skull (↑); made up of a number of fused bones that enclose and protect the brain.

mandible (*n*) lower jaw (↓) bone(s) of a vertebrate.

jaw (*n*) structures of bone and cartilage forming upper and lower parts of the mouth of some animals, e.g. man. They are articulated to provide movement of teeth for biting.

frontal bone (*n*) large bone of a vertebrate skull (↑) covering the front of the brain.

occiput (*n*) (1) region at the back of a vertebrate skull (↑) where it joins the backbone; (2) bony plate at the back of the head in insects.

occipital condyle (*n*) knob of bone on the occiput (↑) at the back of a vertebrate skull (↑) where it articulates with the atlas (↑) to allow the head to move up and down.

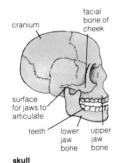

cranium

facial bone of cheek

surface for jaws to articulate

teeth · lower jaw bone · upper jaw bone

skull
human skull

sternum, breastbone (*n*) rod-shaped bone in the middle of the ventral side of the chest in many vertebrates. It articulates with the pectoral girdle and ribs.

rib (*n*) a flattened, curved bone in the chest region of the skeleton of many vertebrates. It articulates with the backbone at one end; at the other it may or may not be connected to the sternum (↑). **True ribs** are attached to backbone and sternum; **false ribs** are attached to backbone and to a true rib; **floating ribs** are attached to backbone only, they are free at the other end. Ribs partially encircle and protect the thoracic cavity and are involved in respiratory movements. Ribs are connected by intercostal muscles.

rib cage (*n*) cage formed by ribs (↑), together with sternum and backbone; in mammals it encloses thorax.

pectoral girdles, shoulder girdles (*n*) part of the skeleton that supports the front appendages in vertebrates. In man they support the arms. Each consists of two bones, the scapula (↓) and clavicle (↓). At no point are they fused to the backbone.

scapula (*n*) bone of the shoulder girdle (↑). In mammals, e.g. man, it is the **shoulder blade** and is a flat, triangular-shaped bone that articulates with the humerus and clavicle (↓). Muscles from the arm are attached to it.

clavicle (*n*) bone of the pectoral girdle (↑) of some vertebrates. It articulates with the scapula (↑) and sternum (↑). In man it is the **collar bone**.

pelvic girdles, hip girdles (*n.pl.*) part of the skeleton that supports the hind appendages in vertebrates. In man they support the legs and each consists of the ilium (↓), pubis (↓) and ischium (↓) which are fused together. The ilium articulates with the sacrum to unite the girdle to the axial skeleton. All the bones have muscle attachments.

pelvis (*n*) (1) cavity at the lower end of the trunk surrounded by the pelvic girdles (↑); (2) the pelvic girdles.

ilium (*n*) bone in each pelvic girdle (↑). In amphibians, reptiles, birds, mammals it is fused to one or more sacral vertebrae to give stability. It joins the ischium (↓) and pubis (↓) at the acetabulum (↓). The shape varies in different animals; in man it is dish-shaped.

ischium (*n*) bone in each pelvic girdle (↑). It joins the ilium (↑) and pubis (↓) at the acetabulum (↓). It bears the weight of a primate when it sits.

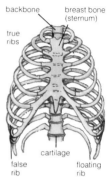

rib cage
human rib cage

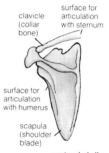

pectoral girdle

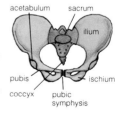

pelvic girdle
human pelvic girdles
forming pelvis

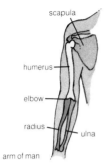

scapula

humerus

elbow

radius

ulna

arm of man

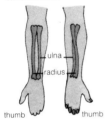

rotation of radius

ulna

radius

thumb   thumb

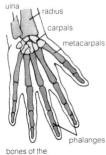

ulna

radius

carpals

metacarpals

phalanges

bones of the
human hand

pubis, pubic bone (n) bone in each pelvic girdle (↑) of amphibians, reptiles, birds, mammals. The right and left pubic bones are sometimes fused. It joins the ilium (↑) and ischium (↑) at the acetabulum (↓).

pubic symphysis (n) joint at which right and left pubic bones (↑) of the pelvic girdles (↑) are fused. Found in mammals and many reptiles.

acetabulum (n) in amphibians, birds, reptiles and mammals a hollow, cup-like socket into which the head of the femur fits to form the hip joint. There is one on each pelvic girdle (↑). At it the ilium, ischium and pubis (↑) join.

humerus (n) long bone found in the upper part of the fore limb between the shoulder and elbow of amphibians, reptiles, birds, mammals. It articulates with the scapula (↑) and with the ulna (↓) and radius (↓).

radius (n) one of the two bones in the lower fore limb of amphibians, reptiles, birds, mammals. Together with another bone, the ulna (↓), it forms the elbow joint with the humerus (↑). It articulates with the fore limb (hand in man) at the thumb side. The radius rotates when the fore limb or hand is turned over.

ulna (n) one of the two bones in the lower fore limb of amphibians, reptiles, birds, mammals. Together with another bone, the radius (↑), it forms the elbow joint with the humerus (↑). It articulates with the fore limb (hand in man) at the side opposite to the thumb.

carpus (n) region of fore limb in amphibians, reptiles, birds, mammals that contains the carpals (↓). In man it is the wrist region.

carpals, carpal bones (n.pl.) a group of several small bones, arranged in two rows, at the carpus (↑). It articulates with the radius (↑) and ulna (↑) and also with the metacarpals (↓).

metacarpals, metacarpal bones (n.pl.) rod-shaped bones of the fore foot of amphibians, birds, reptiles, mammals. In man they form the palm of the hand. One corresponds to each digit. They articulate with the carpals (↑) and phalanges (↓), see diagram.

phalanges (n.pl.) bones in the digits of both fore and hind feet of amphibians, reptiles, birds, mammals. Each digit has 1–5 phalanges joined end to end in a row; each articulates with its neighbours. In man there are three phalanges in fingers and toes and two in the thumb and big toe.

femur (*n*) in amphibians, birds, reptiles, mammals, the long bone found in the upper part of the hind limb, i.e. the thigh. It lies between the hip and knee. It articulates with the hip girdle at the acetabulum (p.91) and with the tibia (↓) at the knee.

tibia, shin bone (*n*) one of the two bones in the lower part of the hind limb of amphibians, reptiles, birds, mammals. It is larger than the other bone, the fibula (↓). It forms the knee joint with the femur (↑) and articulates with the hind foot.

fibula (*n*) outer and smaller of the two bones in the lower part of the hind limb of amphibians, reptiles, birds, mammals. It is situated outside the tibia (↑).

patella (*n*) disc of bone found at the front of the knee joint of most mammals, some birds and reptiles. It protects the joint. It is known as the **knee cap** in man.

tarsus (*n*) region of hind limb in amphibians, reptiles, birds, mammals that contains the tarsals (↓). In man it is in the ankle region.

tarsals, tarsal bones (*n.pl.*) a group of several small bones at the tarsus (↑). In man they form the ankle and heel. They articulate with the tibia (↑) and fibula (↑) and also with the metatarsals (↓).

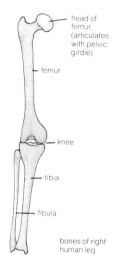

head of femur (articulates with pelvic girdle)

femur

knee

tibia

fibula

bones of right human leg

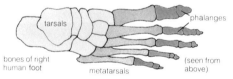

tarsals

phalanges

bones of right human foot

metatarsals

(seen from above)

metatarsals, metatarsal bones (*n.pl.*) rod-shaped bones of the hind foot of amphibians, reptiles, birds, mammals. In man they form the arch of the foot. One corresponds to each digit. They articulate with the tarsals (↑) and phalanges (p.91), *see diagram.*

cartilage (*n*) a hard but flexible tissue in the skeleton of vertebrates. It consists of a matrix in which are scattered rounded cells and fibres. The matrix does not have blood vessels. There are several types of cartilage which vary in the type and density of the fibres, e.g. **white fibro-cartilage** (intevertebral disc), **elastic cartilage** (pinna of the ear), **hyaline cartilage** (wall of the trachea). The type in the skeleton of young children develops into bone; other types remain as cartilage throughout adult life.

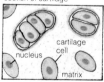

cartilage
section of cartilage

cartilage cell

nucleus

matrix

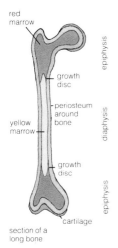

section of a long bone

ossification (*n*) process of changing cartilage into bone. Osteoblasts (↓) start to form a bone matrix at a site called the **centre of ossification**, which is usually in the middle of a bone. The process eventually spreads to both ends of the bone.

growth disc (*n*) region of cartilage in a long bone which allows for growth in length.

osteoblast (*n*) cell found in growing bones. It secretes material to form the bone matrix and eventually becomes trapped in this matrix.

diaphysis (*n*) the shaft of a long limb bone, or the central part of a mammalian vertebra. It contains a centre of ossification (↑) which extends until all the cartilage has been converted into bone. *Compare epiphysis* (↓).

epiphysis (*n*) an end portion of a growing bone that forms an articulating surface of a joint. It is found in mammalian limb bones and in vertebrae. It is ossified separately from the diaphysis (↑). The epiphysis and diaphysis of a growing bone are separated by a plate of cartilage; when growth is complete they fuse.

bone marrow (*n*) connective tissue found inside mammalian bones. **Yellow marrow** contains a high proportion of fatty tissue; it stores fat and makes blood cells. **Red marrow** is found in spongy bone; red blood cells are made in it.

Haversian canals (*n.pl.*) channels in bone that carry blood vessels and nerves to bone cells. Lamellae (↓) and bone cells lie concentrically around them.

lamella (*n*) (*pl.lamellae*) a thin plate-like structure, e.g. a thin layer of bone matrix around a Haversian canal (↑). Lamellae are also arranged parallel to bone surface.

lacuna (*n*) (*pl.lacunae*) a small space, e.g. that containing the bone cells inside the bone matrix (lamella) (↑). Lacunae are arranged around a Haversian canal (↑) (*see diagram*) or parallel to the surface of a bone.

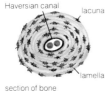

section of bone

periosteum (*n*) layer of connective tissue that surrounds and protects vertebrate bones. It fits very tightly around them; muscles and tendons are attached to it. It is important in bone reconstruction after fracture. The periosteum has a rich blood supply and thus provides nourishment for the bone.

perichondrium (*n*) connective tissue layer surrounding a cartilage. Source of the fibroblasts that form new cartilage by cell division and secretion of matrix and fibres.

joint (*n*) (1) place where two or more separate bones in the skeleton of an animal are united. Joints are classified by the amount of movement they allow; this varies according to the structure of the joint (see fixed joint, gliding joint, hinge joint, ball and socket joint (↓)). Non-fixed joints have cartilage covering the articulating surfaces; it reduces friction and absorbs shock. The bones in movable joints are separated by a fluid-filled cavity; ligaments are attached to both bones and form a fibrous synovial capsule (↓) round the joint. Fixed joints and those allowing limited movement are united directly by cartilage or connective tissue, sometimes by both; (2) place at which separate body parts of invertebrates, particularly of arthropods, meet; (3) alternative term for plant node.

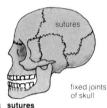

fixed joints of skull

sutures

fixed joint, immovable joint (*n*) joint (↑) which does not allow any movement, e.g. the sutures (↓) which join the bones of the cranium.

sutures (*n.pl.*) (1) immovable joints (↑) between two struc-tures, usually bones, in an animal, e.g. in vertebrate skull between bones of cranium, where their irregular edges are fused together; (2) in plants a line of dehiscence or a line joining two structures, e.g. carpels.

gliding joints of wrist

gliding joint

gliding joint, sliding joint (*n*) joint (↑) allowing limited movement; the two bone surfaces slide over each other, e.g. as between vertebrae.

hinge joint (*n*) joint (↑) which allows movement in one plane only, e.g. knee joint between femur and tibia; elbow joint between humerus and both ulna and radius. The convex end of one bone articulates with the concave end of the other bone.

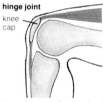

hinge joint

knee cap

hinge joint of knee (side view)

ball and socket joint (*n*) joint (↑) which allows movement in three dimensions, e.g. hip joint between femur and pelvic girdle; shoulder joint between humerus and scapula. A rounded ball-shaped end of one bone fits into a cup-like hollow or socket on the other bone.

condyle (*n*) curved, convex surface of a bone forming a joint (↑). It allows movement in two dimensions but only allows limited rotation, e.g. condyles of femur and tibia at the knee; occipital condyle where the skull articulates with the atlas; condyle at each side of the lower jaw where it articulates with the skull.

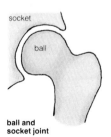

socket

ball

ligament (*n*) band of connective tissue between the two bones of a joint. It prevents dislocation of the joint and restricts movement to certain directions.

ball and socket joint

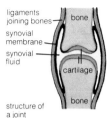

ligaments joining bones
bone
synovial membrane
synovial fluid
cartilage
bone

structure of a joint

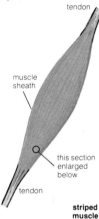

tendon

muscle sheath

this section enlarged below

tendon

striped muscle

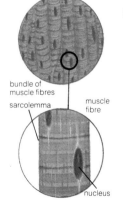

bundle of muscle fibres

sarcolemma

muscle fibre

nucleus

synovial capsule (*n*) a capsule enclosing a freely movable joint, e.g. elbow, knee. Synovial fluid (↓), secreted by the synovial membrane (↓) of the capsule, lubricates the joint.

synovial membrane (*n*) inner layer of a synovial capsule (↑) in a freely movable joint, e.g. elbow, knee. The membrane secretes synovial fluid (↓).

synovial fluid (*n*) viscous fluid secreted by the synovial membrane (↑) of a freely movable joint, e.g. elbow, knee. It nourishes and lubricates the cartilage covering the articulating surfaces.

dislocation (*n*) a dislocated joint. This occurs when the articulating surfaces of the joint are no longer in contact. Ligaments (↑) help prevent this.

sprain (*n*) injury to the ligaments (↑) surrounding a joint.

rheumatoid arthritis (*n*) a disease of the joints (↑); they become inflamed.

osteoarthritis (*n*) form of arthritis (↑) in which cartilage of joint and adjacent bone are worn away.

muscle, muscular tissue (*n*) contractile tissue found in animals. There are three main types: striated muscle (↓), smooth muscle (p.96), cardiac muscle (p.96). This classification is based on appearance under the light microscope and on the type of contraction. When a nerve stimulates a muscle to **contract** the muscular tissue decreases in length and becomes wider. After contraction the muscular tissue **relaxes**. Its return to its normal length and width is usually the result of contraction of antagonistic muscles (p.97).

striped, striated, skeletal, voluntary muscle (*n*) all terms for a type of muscle tissue in vertebrates that is composed of bundles of elongated cells called muscle fibres contained within a muscle sheath. Each fibre is covered with a thin membrane, the sarcolemma (p.96), has many nuclei and contains cytoplasm with longitudinal fibrils which have alternating bands of light and dark coloured tissue. This gives the muscle its striped or striated appearance. Striped muscles are usually attached to bones and move skeletal parts; hence the name skeletal muscle. They are controlled by the brain; hence they are called voluntary muscle. Striped muscles can contract very rapidly and powerfully but only for short periods of time; then they show fatigue (p.96). *Compare smooth muscle, cardiac muscle* (p.96).

muscle fibre (*n*) the elongated cell found in striped muscle (p.95).

sarcolemma (*n*) thin membrane sheath surrounding a muscle fibre (↑).

tendon (*n*) cord of non-elastic tissue that attaches muscle (p.95) to bone. A tough sheath of connective tissue encloses the muscle and continues from the end to form the tendon.

smooth, involuntary, plain, unstriped, unstriated muscle (*n*) all terms for a type of muscle found in vertebrates. It consists of spindle-shaped cells with one central nucleus and no cross striations. Cells are bound tightly together by connective tissue forming sheets that envelop hollow organs, e.g. intestine, bladder, blood vessels. Contraction is controlled by the autonomic nervous system; hence it is called involuntary muscle. Contraction is slow but it can be maintained for long periods without showing fatigue (↓). *Compare striped muscle* (p.95), *cardiac muscle* (↓).

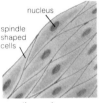

smooth muscle

muscle cell (*n*) spindle-shaped cell found in smooth muscle (p.95). It has one, central nucleus (*see diagram*). Its slow rate of contraction can be maintained for long periods.

cardiac muscle (*n*) type of muscle (p.95) found only in the wall of vertebrate hearts. Its muscle fibres are similar in structure to those of striped muscle (p.95) but they do not have a sarcolemma (↑). They form a network of fibres which are thus held firmly together so that the heart walls can withstand the rhythmic contractions of the heart beating. Heart muscle shows autonomic, rhythmic contraction. It contracts faster than involuntary muscle (↑) but slower than voluntary muscle (p.95). It is controlled by the pacemaker.

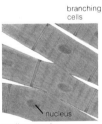

cardiac muscle

muscle tone, – tonus (*n*) state of a muscle showing steady, partial contraction to maintain the shape of an organ. This is brought about by continuous nervous stimulation. Muscle tone is responsible for controlling the posture (p.98) of the body.

muscle fatigue (*n*) tiredness in muscles. It is caused by the accumulation of excess lactic acid in the muscle fibres of voluntary muscles during prolonged exercise. Smooth and heart muscle do not become fatigued under normal circumstances.

muscle cramp (*n*) an involuntary, painful contraction of one or more voluntary muscles.

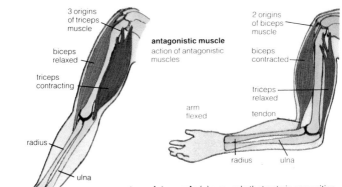

3 origins of triceps muscle

biceps relaxed

triceps contracting

radius

ulna

arm extended

antagonistic muscle
action of antagonistic muscles

2 origins of biceps muscle

biceps contracted

triceps relaxed

arm flexed

tendon

radius   ulna

antagonist muscle (*n*) a muscle that acts in opposition to another muscle, or sometimes several muscles, to bring about the opposing action. All muscles at articulating joints are antagonist muscles since muscles can only contract or relax, not lengthen. For example, the triceps in the arm is the antagonist of the biceps and the biceps is the antagonist of the triceps. These muscles work together to move the arm. The biceps, the flexor (↓), contracts and raises the forearm; at the same time the triceps relaxes. When the triceps, the extensor (↓), contracts it straightens the arm; at the same time the biceps relaxes.

flexor muscle (*n*) a muscle that bends a limb when it contracts, e.g. biceps muscle of the arm is a flexor muscle that bends the arm when it contracts and pulls the forearm towards the upper arm.

flexion (*n*) the action of a flexor muscle (↑). **flex** (*v*).

extensor muscle (*n*) a muscle that extends or straightens a limb when it contracts, e.g. triceps muscle of the arm is an extensor muscle that straightens the arm when it contracts.

extension (*n*) the action of an extensor muscle (↑).

extend (*v*) to enlarge, expand, stretch or straighten out.

biceps (*n*) a muscle with two origins (p.85), e.g. the flexor muscle (↑) of the arm.

triceps (*n*) a muscle with three origins (p.85), e.g. the extensor muscle (↑) of the arm.

quadriceps (*n*) a muscle with four origins (p.85), e.g. the muscle at the front of the thigh extending to the lower leg.

actomyosin (*n*) a complex of two proteins, actin and myosin (↓). It is the most important constituent of muscle and is the major constituent of muscle fibres. Muscular contraction involves shortening of the actomyosin fibrils. This produces heat which helps maintain body temperature.

actin (*n*) protein found in muscle; together with myosin (↓) it forms the complex actomyosin (↑).

myosin (*n*) protein found in muscle; together with actin (↑) it forms the complex actomyosin (↑).

posture (*n*) the way the human body is held by the muscles attached to the skeleton. In good posture the backbone is straight and the weight of the body is balanced through the pelvis on the feet.

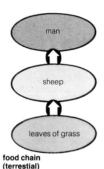

**food chain
(terrestial)**

food chain (*n*) a series of organisms in any natural community in which each member feeds on the one before and is in turn eaten by the one after. Plants and a few bacteria at the bottom of the chain obtain energy from the sun. Plants are eaten by herbivores (p.100); carnivores (p.100) then eat the herbivores. Each link is a trophic level (↓). Organisms at the base of the chain are smaller but more numerous; those at the top are larger but less numerous. There may be one or more species of carnivore at the top of the chain; there are usually several plants or animals at each lower level. Thus in nature the chains usually occur as complex food webs (↓).

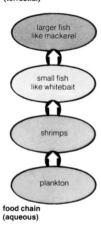

**food chain
(aqueous)**

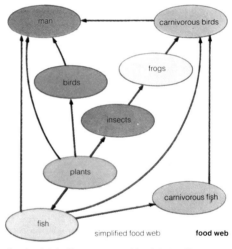

simplified food web   **food web**

pyramid of numbers

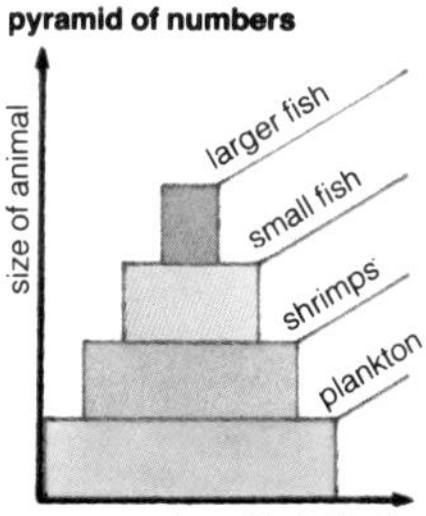

food web (*n*) all interconnected food chains (↑) in a natural community.

trophic level (*n*) one stage in a food chain (↑). Plants are at the first trophic level, herbivores (p.100) that eat them form the second level. Carnivores (p.100) are at succeeding higher trophic levels.

pyramid of numbers (*n*) diagrammatic representation of the numbers of animals at various trophic levels (↑) in a food chain (↑). The numbers are numerically less at each succeeding level but the animals are normally larger than those at lower levels.

herbivore (*n*) a plant-eating animal. *Compare carnivore* (↓), *omnivore* (↓).

ruminant (*n*) a herbivorous (↑) animal which has 3–4 chambers in its stomach. Food passes directly into the first chamber; it is later regurgitated for chewing.

saprophyte (*n*) organism which lives on and obtains its nutrients from decaying organic matter, e.g. many fungi and bacteria. *Compare herbivore* (↑), *omnivore* (↓).

carnivore (*n*) a flesh-eating animal. *Compare herbivore* (↑), *omnivore* (↓).

predator (*n*) an animal that kills other animals (its prey) for food. Carnivores (↑) are predators.

insectivore (*n*) animal that eats insects, e.g. shrew, mole.

saprophyte — a fungus obtaining nutrients from decaying leaves and wood

insectivore

shrew

scavanger (*n*) animal that feeds on remains left by carnivores (↑).

omnivore (*n*) an animal that eats both plants and animals. *Compare herbivore* (↑), *carnivore* (↑).

autotrophic (*adj*) describes organisms such as a few bacteria and all green plants that do not depend on outside sources to supply organic substances as nutrients. They build up food materials from inorganic compounds using energy also obtained independently of organic matter. Plants make organic compounds by photosynthesis. All other organisms are heterotrophic (↓).

heterotrophic (*adj*) describes organisms that must obtain their organic food from the environment, e.g. all animals, fungi, most bacteria. They can be herbivores, carnivores, saprophytes (↑) or parasites (p.210); all their organic material can be traced back to the activity of autotrophic (↑) organisms.

holophytic (*adj*) describes organisms that feed like a plant, i.e. they photosynthesise.

holozoic (*adj*) describes organisms that feed like animals, i.e. by ingesting organic substances as solids or liquids.

alimentary canal, gut (*n*) canal or tube in animals into which foodstuffs pass to be broken down, i.e. digested. Digested food is absorbed from it and assimilated by body tissues. In some animals, e.g. flatworms, it has only one opening; in most it has two, the mouth into which food is taken and the anus from which undigested material is expelled. The alimentary canal is adapted to suit the diet of the animal. Succeeding parts of it deal with ingestion, digestion, absorption, egestion.

mouth

structure of the mouth

mouth (*n*) opening to the exterior in the head of animals; food is taken in through it.

lip (*n*) (1) one of the two, soft flaps of tissue in front of the teeth and around the mouth of an animal; (2) edge of an orifice, cavity or vessel; may be bent outwards forming a rim.

buccal cavity (*n*) cavity of the mouth (↑) within the lips; leads to the pharynx. In man it contains the teeth and tongue. The roof of it is formed by the hard palate (↓).

palate (*n*) roof of a vertebrate mouth (↑). In mammals and crocodiles a **false palate** has developed beneath the original palate to separate the alimentary canal (↑) and the nasal passages. The false palate is divided into a bony **hard palate** that forms the roof of the buccal cavity (↑) and the **soft palate** which is raised during swallowing to prevent food entering the nasal cavity.

tonsils (*n*) mass of tissue in the mouth (↑) of many vertebrates. It is found at the base of the tongue (↓) and at the entrance to the pharynx.

uvula (*n*) fleshy, conical mass of tissue suspended from the palate (↑) over the back part of the tongue (↓).

tongue (*n*) fleshy, muscular organ attached to the floor of the buccal cavity (↑) of a vertebrate. It is used for tasting and to help masticate and swallow food. In man it is also used in speech.

lingual (*adj*) concerned with the tongue (↑) and speech.

salivary glands (*n*) glands that secrete saliva (p.111) into the buccal cavity (↑) of vertebrates, or foregut (oesophagus) of other animals, e.g. insects.

glottis (*n*) opening of the trachea into the pharynx of vertebrates. In mammals vocal cords are stretched across the opening.

epiglottis (*n*) flap of cartilage and membrane above the glottis (↑). The glottis is pushed against it during swallowing to prevent food entering the trachea.

tooth (*n*) (*pl.teeth*) a small, hard structure found in the mouth of vertebrates. In fish, amphibians and reptiles they are distributed over the palate. In crocodiles and mammals they are less numerous and are found in the jaws. A tooth is divided into a crown (↓) which is covered in enamel and a root (↓) which lies inside the gum (↓). Teeth are used for fighting, seizing and tearing prey, masticating food. Mammals have four types, incisors (↓), canines (↓), premolars (↓) and molars (↓). The number and size varies from species to species. Most mammals, including man, have milk teeth (↓) in infancy that are replaced later by permanent teeth (↓).

milk teeth, deciduous teeth (*n*) the first set of teeth (↑) in most mammals, including man. They consist of some or all of incisors (↓), canines (↓), premolars (↓). They are fewer in number than permanent teeth (↓) by which they are later replaced.

permanent teeth (*n*) second set of teeth (↑) in mammals; they replace milk teeth (↑). They consist of incisors (↓), canines (↓), premolars (↓), molars (↓).

incisor (*n*) sharp, chisel-shaped mammalian tooth (↑) found at front of the mouth. It has one root (↓) and is used for gnawing, nibbling, biting, cutting. Those in rodents grow in length continuously.

canine, dog or **eye tooth** (*n*) sharp, pointed, conical mammalian tooth (↑). One is usually found on either side of the lower and upper jaws between the incisors (↑) and premolars (↓). Each has one root (↓) and is used for holding prey and tearing or biting flesh. They are well developed in carnivores, practically absent in rodents, but may be well developed in some herbivores or omnivores, for defence, e.g. baboon.

premolar (*n*) mammalian tooth (↑) with a flat, ridged surface. Two or more are found between the canines (↑) (or incisors (↑) if canines are absent) and molars (↓). Each has two roots (↓) and is used to crush and grind food.

molar (*n*) mammalian tooth (↑) similar to a premolar (↑) in shape and function but differing in having two or more roots (↓) and being present only in permanent teeth (↑). Found in the back of the mouth behind the premolars.

wisdom tooth (*n*) the back four molars (↑) in man; they appear last, usually in late teens.

carnassial tooth (*n*) molar or premolar tooth characteristic of carnivores; they have a shearing action.

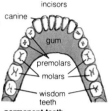

permanent teeth
lower jaw of adult man

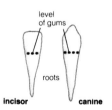

incisor   **canine**

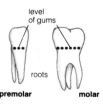

premolar   **molar**

types of teeth

structure of a tooth

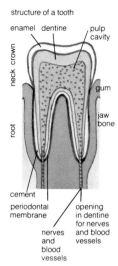

impacted tooth (*n*) tooth wedged between jaw bone and another tooth so that it cannot come through the gum.

crown (*n*) the top part of a structure, e.g. the upper part of a tooth (↑) which is exposed above the gum (↓).

root (*n*) the base or growing point of a structure, e.g. the part of a mammalian tooth (↑) which lies within the jaw bone. Incisors (↑) and canines (↑) have one root, premolars (↑) two, molars (↑) two or more.

neck (*n*) narrowed part of a structure, e.g. the portion of a tooth between crown (↑) and root (↑).

cusp (*n*) projection on the biting surface of a mammalian tooth.

gum (*n*) soft tissue around roots of mammalian teeth.

cement (*n*) hard, bone-like substance covering the dentine (↓) in the root portion of a mammalian tooth.

enamel (*n*) a hard, white covering over the dentine (↓) on the crown (↑) of a tooth. It is almost entirely composed of inorganic material (mainly calcium salts).

dentine (*n*) a hard, yellow material that makes up most of the structure of a tooth. It is similar chemically to bone but contains no cells.

pulp cavity (*n*) cavity inside a tooth which contains dental pulp (↓). The cavity is surrounded by dentine (↑) except for a small canal at the base of each root which allows nerves and blood vessels to pass into the pulp.

dental pulp (*n*) mass of soft, vascular tissue in the pulp cavity (↑) of a tooth. Odontoblasts (↓) are found round its edges.

odontoblast (*n*) cell found round the edge of dental pulp (↑). In growing teeth it forms dentine (↑).

periodontal membrane (*n*) thin fibrous membrane surrounding the root (↑) of a mammalian tooth; binds tooth into its socket.

dentistry (*n*) study of teeth and their care. The person carrying this out is called a **dentist**.

caries (*n*) decay, especially of the teeth. This should be treated by a dentist (↑).

plaque (*n*) in dentistry (↑), a film of saliva and bacteria that forms on teeth.

tartar (*n*) deposit on teeth, mainly calcium phosphate.

dentition (*n*) (1) the number, type and arrangement of teeth (↑) in an animal. Most mammals have heterodont dentition (p.104); most non-mammalian vertebrates have homodont dentition (p.104). (2) the growth of teeth and their emergence from the gums.

carnassial tooth

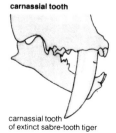

carnassial tooth
of extinct sabre-tooth tiger

homodont dentition (*n*) dentition (p.103) of most non-mammalian vertebrates. The teeth are all of the same type and function.

heterodont dentition (*n*) dentition (p.103) of most mammals. The teeth are of various kinds, each having a particular function.

dental formula (*n*) way of expressing, in the form of fractions, the number of each type of tooth in a mammal. The numbers above the line show the number of each type of tooth in half the upper jaw, the numbers below give similar information for the bottom jaw. Each fraction is preceded by i,c,p (or pm), m, the initial letter of the type of tooth.

man
$i\frac{2}{2}$ $c\frac{1}{1}$ $pm\frac{2}{2}$ $m\frac{3}{3}$

rabbit
$i\frac{2}{1}$ $c\frac{0}{0}$ $pm\frac{3}{2}$ $m\frac{3}{3}$

key
i   incisors
c   canines
pm premolars
m molars

dental formula

pharynx (*n*) a thick-walled, muscular tube of the alimentary canal lying between mouth and oesophagus (↓). In some amphibians, reptiles, birds and mammals, the glottis opens into it. In fish and aquatic amphibians gill slits arise from it. In mammals like man the throat and back of the nose connect with it and it is partially divided by the soft palate *(see p.101)*. Stimulation of the pharynx by food causes swallowing.

throat (*n*) region of the body that lies at the front of the neck and contains the oesophagus (↓) and trachea.

oesophagus, gullet (*n*) a muscular tube that runs from pharynx (↑) to stomach (↓) in vertebrates. In birds it contains the crop (p.44). It passes food by peristalsis to the digestive parts of the alimentary canal.

stomach (*n*) a J- or U-shaped structure of the alimentary canal. In vertebrates, except birds, it follows the oesophagus (↑) and has thick, muscular walls to churn and so mix food. Secretory cells lining its walls secrete acidic gastric (↓) juices which carry out some digestive processes. Food passes from the stomach to the intestines (p.106) via the pylorus (↓). The stomach is often in separate sections, e.g. ruminants have several chambers (see rumen (↓)), birds have a crop (p.44) and gizzard (↓).

gizzard (*n*) in birds and most invertebrates, a strong, muscular-walled sac of the alimentary canal; found immediately after the crop (p.44). Food is crushed and mixed in it.

rumen (*n*) first chamber of a ruminant stomach. It is used mainly for temporary storage of food but some cellulose digestion, vitamin B synthesis and sugar absorption also occurs in it.

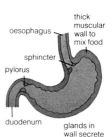

oesophagus — thick muscular wall to mix food

sphincter

pylorus

duodenum — glands in wall secrete gastric juice

stomach
stomach of man

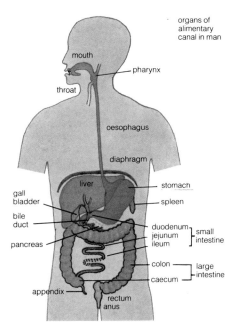

organs of
alimentary
canal in man

mouth

pharynx

throat

oesophagus

diaphragm

liver

gall
bladder

bile
duct

pancreas

appendix

stomach

spleen

duodenum
jejunum   small
ileum   intestine

colon   large
              intestine
caecum

rectum
anus

gastric (*adj*) concerned with the stomach or the
digestive processes.

pylorus, pyloric orifice (*n*) constriction at the end of the
stomach (↑) where it leads into the intestines (p.106). It
can be closed in many animals by the sphincter muscle
(↓); this prevents food from leaving the stomach before
the digestive processes are completed.

pyloric sphincter (*n*) ring of strong muscles round the
pylorus (↑) which can close the stomach off from the
intestines (p.106).

peritoneum (*n*) membrane lining the peritoneal cavity. It
forms the mesentery (↓) which surrounds and supports
the viscera (i.e. stomach, intestines, spleen, etc.).

mesentery (*n*) double layer of peritoneum (↑) that
attaches the viscera (i.e. stomach, intestines, spleen,
etc.) to the wall of the peritoneal cavity. Mesentery
contains the blood vessels, lymph vessels and nerves
which supply these organs. Sometimes the term is used
for tissue supporting the small intestine (p.106) only.

intestine (*n*) long, tube-like part of the alimentary canal; concerned with digestion and absorption of food and the formation of faeces. In vertebrates it lies between the stomach and rectum. In many animals the internal surface area is increased by folding and the presence of projections (e.g. villi (↓) in man) on the inner wall. The intestine is very long (about 3½ m in man) and therefore coiled within the abdominal cavity. Its muscular walls pass the contents to the rectum by peristalsis. The intestine may be specialised into small (↓) and large intestines (↓).

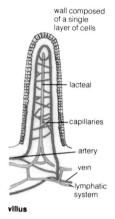

villus

small intestine (*n*) specialised first part of intestine (↑) in reptiles, birds, mammals. It is narrower and usually longer than the second part, the large intestine (↓). Secretions from glands in its wall and from the pancreas and liver complete food digestion in it (except herbivores). The internal surface area is increased by villi (↓) into which digestion products are absorbed. Amino acids and monosaccharides pass from it into blood vessels; fats pass into lacteals. *See duodenum* (↓), *jejunum* (↓), *ileum* (↓).

duodenum (*n*) first part of the small intestine (↑). It leads directly into the jejunum (↓). The bile duct and pancreatic duct open into it. Secretions from these together with those from the walls digest food.

jejunum (*n*) wider, middle part of the small intestine (↑) between duodenum (↑) and ileum (↓). It has larger villi (↓) than the rest of the small intestine and is the main region for food absorption.

ileum (*n*) last part of the small intestine (↑) after the jejunum (↑). It leads into the large intestine (↓). Absorption of digested food is completed here. The term can refer collectively to the ileum and jejunum.

villus (*n*) (*pl. villi*) any finger-like projection, e.g. those on the wall lining the small intestine. Each contains muscle to move it and a lacteal surrounded by capillaries into which all digested food is absorbed. Villi greatly increase the surface area of the intestine.

large intestine (*n*) second part of intestine (↑) in reptiles, birds and mammals. It is wider and shorter than the small intestine (↑). It absorbs water from undigested food and thus forms the faeces. *See colon* (↓), *caecum* (↓)

colon (*n*) first part of large intestine (↑). It leads directly into the narrower rectum (↓). Water is absorbed from the undigested part of food in it.

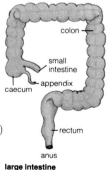

large intestine

caecum (n) a blind-ended branch of a hollow organ, particularly that of the large intestine (↑). Some herbivores have a wide caecum; cellulose is digested in it. In man it is much reduced and has no function.

appendix, vermiform appendix (n) small, blind-ending tube. It is found at the end of the caecum (↑).

appendicitis (n) inflammation of the appendix (↑).

rectum (n) second part of the large intestine (↑). It is a short tube which stores faeces (p.108) and expels them through the anus (↓) or cloaca (p.155).

anus (n) terminal opening of the alimentary canal through which faeces (p.108) are expelled. It normally opens from the rectum (↑). It is closed by the **anal sphincter**. *Compare cloaca* (p.155).

pancreas (n) gland found, supported by mesentery, near the duodenum of vertebrates. It produces (1) an alkaline mixture of enzymes, mainly trypsinogen, lipase, amylase, maltase, which pass via the pancreatic duct to the duodenum; (2) the hormone insulin from a group of cells, the islets of Langerhans (see p.183).

bile, gall (n) secretion of vertebrate liver (p.130). It is stored in the gall bladder and passes via the bile duct to the duodenum. It is a bitter, alkaline liquid consisting mainly of **bile salts** and **bile pigments**. Bile salts lower the surface tension of chyme (↓) and emulsify fats before their enzymic digestion. Bile pigments are waste products of haemoglobin destruction. Also, toxic substances are eliminated from the body in bile.

emulsify (v) to suspend one liquid as small drops in another liquid. **emulsification, emulsion** (ns).

pancreatic duct (n) duct originating from the pancreas (↑); joined by the bile duct (↓) before it enters duodenum.

bile duct (n) duct carrying bile from the vertebrate liver. It joins the pancreatic duct (↑).

peristalsis (n) waves of contraction that pass along the smooth muscles enclosing tubular organs, particularly the intestines. It mixes the contents and passes them from one end of the organ to the other.

bolus (n) any round mass of material, e.g. a food ball passing down the oesophagus by peristalsis (↑).

chyme (n) soft, milky mass of partially digested food as it leaves the vertebrate stomach.

chyle (n) liquid formed in lacteals of villi as a result of absorption of fat. It is a fine emulsion (↑) of fat in lymph.

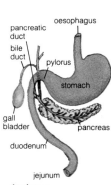

pancreatic duct
bile duct
oesophagus
pylorus
stomach
gall bladder
pancreas
duodenum
jejunum

duodenum
duodenum and pancreas of man

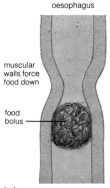

oesophagus
muscular walls force food down
food bolus

bolus

faeces (*n.pl.*) residue of undigested food, bile, other secretions and bacteria which is expelled from the anus in mammals and the cloaca in other vertebrates.

egest (*v*) to get rid of undigested food, e.g. as faeces.

defaecate (*v*) to expel faeces (↑) from the anus or cloaca.

ingest (*v*) to take in solid material for use as food.

digestion (*n*) breakdown of complex foodstuffs into simple molecules which can be absorbed by an organism for metabolic purposes. In most animals it is carried out by enzymes in the alimentary canal.

absorption (*n*) any process in which soluble substances in solution pass through a membrane into an organism or part of it. *Compare adsorption* (↓).

adsorption (*n*) the formation of a layer of a substance on the surface of a solid or liquid. *Contrast absorption* (↑) where the substance penetrates into the interior.

assimilation (*n*) process of utilisation of simple molecules, e.g. amino acids into complex compounds like proteins.

metabolism (*n*) sum of all the chemical processes taking place in an organism or parts of it. The processes mainly involve both breakdown of complex organic molecules to simpler molecules, combined with the synthesis of ATP, i.e. catabolism (↓) and synthesis of complex organic molecules from simple molecules in reactions which utilise ATP, i.e. anabolism (↓). The term is also used for metabolic changes of a constituent of an organism, e.g. carbohydrate metabolism.

metabolite (*n*) chemical substance used in metabolism (↑); either made by an organism or taken from the environment.

anabolism, biosynthesis (*n*) production of chemical compounds by living organisms from simple precursors (e.g. elements, carbon dioxide, water) and their use in the synthesis of larger molecules. ATP supplies energy needed. *See metabolism* (↑), *catabolism* (↓).

catabolism (*n*) breakdown of complex molecules (e.g. carbohydrates) into simple molecules (e.g. sugars) and also the further breakdown, e.g. to carbon dioxide and water. The energy released is stored as ATP and used for anabolism (↑).

basal metabolism (*n*) metabolism (↑) of an animal at rest. It is the energy needed to maintain vital processes like circulation, respiration, muscle tone, but excludes that needed for movement, growth, reproduction, etc. Its rate is called the **basal metabolic rate (BMR)**.

substrate molecule on enzyme at active site

enzyme molecule

enzyme
active site of enzyme

enzyme (*n*) complex protein produced by living cells of bacteria, plants and animals. It acts as a catalyst (↓) in biological reactions converting one or more substances, the substrates (↓) into other substances, the products. After the reaction it is unchanged and ready to catalyse the change of more substrates. Very small amounts of enzymes are needed. Enzymes are named by the substrate and by the type of reaction they catalyse, e.g. alcohol dehydrogenase catalyses removal of hydrogen atoms from ethanol. Most are highly specific both to substrate and type of reaction they catalyse. Each needs specific conditions for optimum activity, e.g. pH, presence of coenzymes (↓) and activators, absence of inhibitors (p.110). Most work best at a temperature of 35–40°C. Being protein, many are destroyed above 50°C and easily inactivated chemically.

catalyst (*n*) a substance that increases the rate of a chemical reaction without itself being changed permanently or used up in the reaction. Does not alter the products of that reaction. They are needed in small amounts and require specific conditions for optimal activity. Enzymes (↑) are biological catalysts.

substrate (*n*) substance on which an enzyme (↑) acts to form a product(s), e.g. starch is substrate for amylase.

active centre, -site (*n*) the small part of an enzyme (↑) molecule that combines with the substrate (↑) and changes it into the product.

cofactor (*n*) substance necessary for the activity of some enzymes. They can be metal ions, e.g. Na^+, K^+, Mg^{2+} or organic molecules called coenzymes (↓).

coenzyme (*n*) an organic cofactor (↑) necessary for the activity of some enzymes. There are many coenzymes, e.g. NAD (↓), NADP (↓), FAD (p.110), FMN (p.110). One coenzyme may function in reactions catalysed by many different enzymes. Many vitamins, e.g. riboflavin (p.127) are coenzymes.

NAD (nicotinamide adenine dinucleotide) (*n*) a coenzyme (↑) in many oxidation-reduction reactions. It was formerly known as DPN or coenzyme I.

NADH (*n*) the reduced form of NAD (↑).

NADP (nicotinamide adenine dinucleotide phosphate) (*n*) a coenzyme (↑) in many oxidation-reduction reactions. It was formerly known as TPN or coenzyme II.

NADPH (*n*) the reduced form of NADP (↑).

NAD

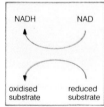

NADH   NAD

oxidised substrate   reduced substrate

action of NAD

NADP

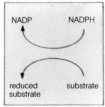

NADP   NADPH

reduced substrate   substrate

action of NADP

FAD (flavin adenine dinucleotide) (*n*) a coenzyme (p.109) in some oxidation-reduction reactions.

FMN (flavin mononucleotide) (*n*) a coenzyme (p.109) in some oxidation-reduction reactions.

enzyme kinetics (*n*) study of the effects of factors such as substrate concentration, temperature or pH, on the rates of enzyme (p.109) reactions.

Q_{10}, temperature coefficient (*n*) the ratio by which the velocity of an enzyme reaction increases for a rise in temperature of 10°C.

enzyme inhibitor (*n*) substance that reduces the rate of an enzyme reaction. Inhibition can be irreversible or reversible.

feedback inhibition (*n*) the inhibition of an enzyme (p.109) by the product of its reaction.

isomerase (*n*) one of a group of enzymes (p.109) where the product is an isomer of the substrate.

hydrolase (*n*) enzyme (p.109) that catalyses breakdown of a substrate by addition of a water molecule.

dehydrogenase (*n*) enzyme (p.109) that catalyses oxidation of a substrate by removal of hydrogen. The hydrogen usually combines with a coenzyme such as NAD (p.109). Most biological oxidations are carried out by dehydrogenases. *Compare oxidase* (↓).

oxidase (*n*) a type of dehydrogenase (↑) in which the hydrogen removed is combined with molecular oxygen.

peroxidase (*n*) a special type of oxidase (↑) in which the hydrogen removed is transferred to hydrogen peroxide (H_2O_2) to form water.

catalase (*n*) a peroxidase (↑) enzyme that decomposes hydrogen peroxide to molecular oxygen and water.

transferase (*n*) one of a group of enzymes (p.109) that catalyse the transfer of a group of atoms, other than hydrogen, from a molecule of one substance to a molecule of another substance, e.g. **transaminase** catalyses the transfer of an amino group, **carboxylase** that of a carboxyl group.

kinase (*n*) one of a group of enzymes (p.109) that transfer the terminal phosphate of ATP (↓) to the substrate.

ATP (adenosine triphosphate) (*n*) the source of energy for many enzyme reactions. The energy is stored in the terminal two phosphate groups. In its simplest reactions the terminal phosphate group is transferred to another substance and **ADP (adenosine diphosphate)** is formed. ATP is regenerated from ADP in the mitochondria.

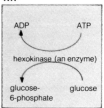

ATP

ADP   ATP

hexokinase (an enzyme)

glucose-6-phosphate   glucose

action of ATP

ribose — adenine
|
phosphate
}
phosphate
}
phosphate

} energy-rich bond

ATP molecule showing energy-rich bonds

PART OF BODY   ENZYME

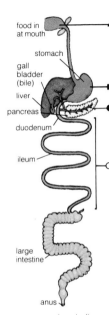

food in at mouth ▲

stomach

gall bladder (bile)

liver

pancreas

duodenum

ileum

large intestine

anus

▲ ptyalin
  pepsin
■ rennin
  lipase
  amylase
● maltase
  trypsinogen
  enterokinase
  trypsin
  erepsin
○ maltase
  sucrase
  lactase

summary of digestive enzymes

cellulase (n) an enzyme (p.109) found in bacteria, fungi and some insects which hydrolyses cellulose to simpler carbohydrates.

zymase (n) an enzyme system found in yeasts in which several enzymes act to convert hexose sugars (e.g. glucose) into ethanol and carbon dioxide.

ptyalin (n) an enzyme with amylase-like (p.112) activity found in saliva (↓) of some mammals, e.g. man. It hydrolyses starch or glycogen to maltose (p.117).

saliva (n) secretion from the salivary glands. In terrestrial vertebrates it contains mucus to moisten and lubricate food for swallowing; in addition, in some, ptyalin (↑) is present. Saliva is secreted on to the mouth parts of insects. In some (e.g. housefly) it contains digestive enzymes, in others (e.g. bloodsuckers) an anticoagulant.

gastric juice (n) secretion from glands in stomach wall of animals. In vertebrates it contains mucus, hydrochloric acid, pepsin (↓) and in young mammals rennin (↓).

juice (n) (1) secretion from a digestive gland, e.g. gastric juice (↑) from the stomach, pancreatic juice (↓) from the pancreas. It is a liquid and contains enzymes. (2) liquid part of plants, e.g. fruit juice obtained by crushing fruit.

proteolytic enzyme, protease (n) an enzyme that hydrolyses proteins. There are two types, proteinases (↓) and peptidases (↓).

proteolysis (n) the hydrolysis of proteins into their constituent amino acids either chemically (e.g. with acids) or by proteolytic enzymes (↑).

proteinase (n) a proteolytic enzyme (↑) that splits protein molecules into smaller fractions called peptides, e.g. pepsin (↓).

peptidase (n) a proteolytic enzyme (↑) that splits peptides into their constituent amino acids, e.g. trypsin (p.112) and chymotrypsin (p.112).

pepsin (n) proteolytic enzyme (↑) found in gastric juice secreted by glands in the wall of a vertebrate stomach. In acid solutions it splits proteins to peptides.

rennin (n) an enzyme secreted in the stomach of young mammals. It clots milk and digests milk protein.

pancreatic juice, pancreatin (n) secretion of the pancreas. It passes to the duodenum via the pancreatic duct. It is made alkaline by bile. Several enzymes are found in it, mainly lipase, amylase, trypsinogen, maltase (p.112).

amylase, diastase (*n*) one of a group of enzymes found in plants and also in saliva (p.111) and pancreatic juice (p.111) of animals. It splits starch or glycogen, mainly into the disaccharide maltose (p.117).

lipase (*n*) an enzyme which hydrolyses fats into glycerol and fatty acids. In vertebrates it is secreted by glands in the small intestine and by the pancreas.

trypsinogen (*n*) the inactive form or zymogen (↓) in which trypsin is secreted by the pancreas. It consists of a trypsin molecule extended by the presence of an additional polypeptide unit. Removal of the polypeptide by enterokinase (↓) and already activated trypsin (↓) converts it into its enzymically active form.

trypsin (*n*) a peptidase (p.111) secreted by the vertebrate pancreas as trypsinogen (↑). Trypsin splits proteins mainly into polypeptides.

enterokinase, enteropeptidase (*n*) an enzyme secreted by glands in the small intestine. It converts trypsinogen (↑) into its active form trypsin (↑).

zymogen (*n*) an inactive enzyme which can be converted by other enzymes into its active form, e.g. trypsinogen (↑).

chymotrypsinogen (*n*) the inactive form in which chymotrypsin (↓) is secreted.

chymotrypsin (*n*) proteolytic enzyme (p.111) secreted by the pancreas as inactive chymotrypsinogen. When activated by trypsin (↑) in the small intestine it splits peptides into their constituent amino acids.

succus entericus, intestinal juice (*n*) juice (p.111) secreted by Brunners glands in the wall of the duodenum of vertebrates. It contains digestive enzymes, e.g. amylase maltase, sucrase, lactase, lipase, proteinases, peptidases.

erepsin (*n*) a mixture of proteolytic enzymes (p.111); mainly splits peptides into amino acids. It is found in the succus entericus (↑) of man.

sucrase, invertase, saccharase (*n*) an enzyme that splits sucrose into glucose and fructose. It is found in man in the succus entericus (↑).

lactase (*n*) an enzyme that splits lactose, i.e. milk sugar into glucose and galactose. It is found in man in the succus entericus (↑).

maltase (*n*) an enzyme that splits maltose into two molecules of glucose. It is found in man in the succus entericus (↑) and in pancreatic juice (p.111).

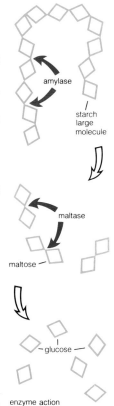

amylase

starch large molecule

maltase

maltose

glucose

enzyme action on starch

carbohydrate (*n*) a group of complex compounds consisting of the elements carbon, hydrogen and oxygen; general formula $C_x(H_2O)_y$. The three main types are monosaccharides ($\downarrow$), e.g. glucose, disaccharides ($\downarrow$), e.g. sucrose and polysaccharides ($\downarrow$), e.g. starch, glycogen, cellulose. Edible carbohydrates include all three types and play an essential role in the metabolism of all organisms as their breakdown yields energy in the form of ATP for biological reactions.

monosaccharides (*n.pl.*) the simplest group of sugars (p.115) with a general formula $C_nH_{2n}O_n$, where n is usually 3–7. They are classified, according to the number of carbon atoms the molecule contains, as trioses (3 atoms), tetroses (4 atoms), pentoses (5 atoms), hexoses (6 atoms). Most monosaccharides in nature are pentoses and hexoses. They are all reducing sugars. Like all sugars they are soluble in water. When decomposed the product does not have the properties of a sugar. *Compare disaccharides* ($\downarrow$).

disaccharides (*n.pl.*) sugars (p.115) formed by the combination of two monosaccharide ($\uparrow$) molecules; have a formula $C_{12}H_{22}O_{11}$. They can be hydrolysed into the constituent monosaccharides. Like all sugars they are soluble in water. The most common disaccharides are maltose, sucrose, lactose.

polysaccharides (*n.pl.*) carbohydrates formed by the combination of a large number (usually the same type) of monosaccharide ($\uparrow$) molecules. They function in plants and animals as storage materials, e.g. starch, glycogen and structural units, e.g. cellulose, chitin. They are usually non-crystalline, insoluble in water and not sweet *(compare sugars* p.115*)*. They form monosaccharides on hydrolysis.

starch (*n*) a carbohydrate ($\uparrow$) with a formula $(C_6H_{10}O_5)_n$. It is an insoluble polysaccharide ($\uparrow$) made up of glucose units linked in a way quite different from that of cellulose (p.114). There are two forms: amylose (p.114) and amylopectin (p.114). Starch stains blue with iodine ($\downarrow$); is formed in green plants by photosynthesis; is hydrolysed by the enzyme amylase to maltose.

iodine test (*n*) test to confirm presence of starch ($\uparrow$). Food containing starch when treated with iodine solution gives a purple or blue-black coloration.

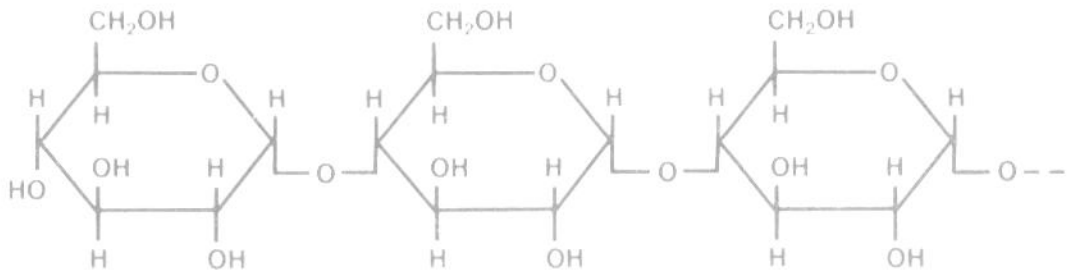

amylose

glucose units
of amylose

amylose (*n*) one of the two forms of starch (p.113). It is
formed from long, unbranched glucose chains.
Contrast amylopectin (↓).

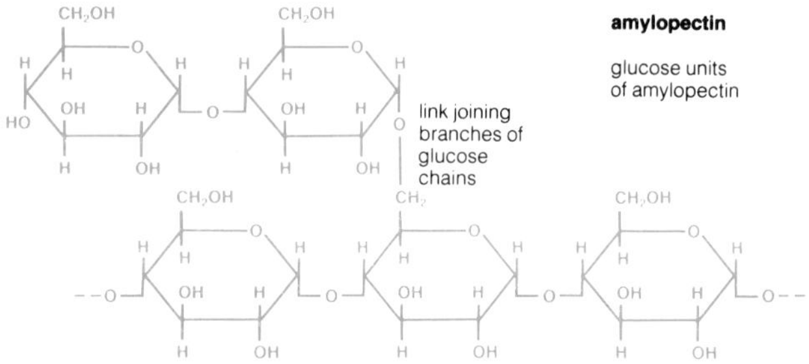

amylopectin

glucose units
of amylopectin

link joining
branches of
glucose
chains

amylopectin (*n*) one of the two forms of starch (p.113). It
is formed from branched glucose chains. *Contrast
amylose* (↑).

glycogen (*n*) an insoluble polysaccharide (p.113)
consisting of branched chains of glucose. It is the form
in which carbohydrate is stored in animals and also in
some algae and fungi. Mammals make and store it
mainly in the liver and muscles.

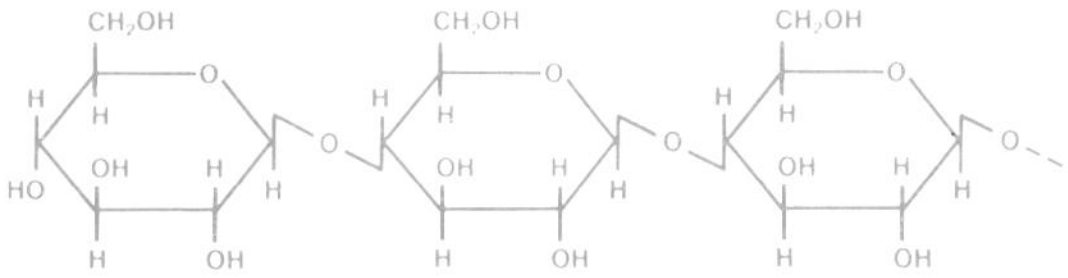

cellulose

glucose units
of cellulose

cellulose (*n*) a complex polysaccharide (p.113)
consisting of long chains of glucose molecules linked in
straight chains unlike those of starch (p.113). It is the
basic constituent of cell walls in algae, some fungi and
all green plants. It has great tensile strength.

dextrin (*n*) a polysaccharide (p.113) consisting of a
short chain of glucose molecules. It may occur as an
intermediate in the hydrolysis of starch to glucose.
Compare dextran (↓).

dextran (*n*) storage polymer of some yeasts and bacteria. It is made up of glucose residues but linked differently from those in dextrin (↑), starch, etc.

chitin (*n*) a nitrogen containing polysaccharide (p.113) that forms a material with great mechanical strength and resistance to chemicals. It is found in the cuticle of insects and the cell walls of many fungi.

pectic compounds (*n.pl.*), **pectin** (*n*) a mixture of acidic polysaccharides (p.113) found in primary cell walls of plants, especially in fruits. Pectins form gels and are used commercially, e.g. in jam manufacture.

sugar (*n*) a white, crystalline carbohydrate (p.113), soluble in water, e.g. glucose, sucrose. Sugars are classified according to their structure as monosaccharides or disaccharides and according to their reaction with Fehling's (↓) and Benedict's (↓) solutions as reducing or non-reducing. All sugars are reducing sugars after hydrolysis.

Fehling's test (*n*) a biochemical test for **reducing sugars**. When warmed with Fehling's solution reducing sugars (all monosaccharides and some disaccharides like maltose and lactose) give a red precipitate of copper (I) oxide. **Non-reducing sugars** (most disaccharides, e.g. sucrose) give a negative result, i.e. the solution remains blue. *See Benedict's test* (↓).

Benedict's test (*n*) a biochemical test for reducing sugars. When warmed with Benedict's solution **reducing sugars** give a red precipitate. **Non-reducing sugars** do not give this positive result. Amounts of sugar that are too small to be detected by the Fehling's test (↑) give a green precipitate with Benedict's solution.

CHO   CHO

H   C   OH   **triose**   H   C   OH   **tetrose**

CH₂OH   H   C   OH

glyceraldehyde,   erythrose,   CH₂OH
a triose sugar   a tetrose sugar

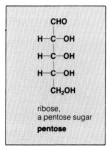

CHO

H—C—OH

H—C—OH

H—C—OH

CH₂OH

ribose,
a pentose sugar
pentose

triose (*n*) a monosaccharide with three carbon atoms, e.g. glyceraldehyde.

tetrose (*n*) a monosaccharide with four carbon atoms, e.g. erythrose.

pentose (*n*) a monosaccharide with five carbon atoms; formula $C_5H_{10}O_5$, e.g. ribose and deoxyribose, important constituents of nucleic acids.

hexose (*n*) a monosaccharide (p.113) with six carbon atoms; formula $C_6H_{12}O_6$, e.g. glucose, fructose, galactose, mannose. Combinations of hexoses make up most of the disaccharides and polysaccharides that are biologically important.

optical activity (*n*) a property that some molecules have of rotating the plane of polarized light. Such molecules have an asymmetric structure, i.e. a carbon atom to which four different atoms or groups of atoms are attached. Monosaccharides have this property. The two configurations, called **optical isomers**, are mirror images, *see diagram*.

```
    CHO                CHO
H   C   OH      HO     C   H
    CH₂OH              CH₂OH
```

optical isomers
two optically active
forms of glyceraldehyde

```
    CHO                CH₂OH              CHO                CHO
H   C   OH         C = O          HO     C   H       H   C   OH
HO  C   H       HO     C   H      HO     C   H       HO  C   H
H   C   OH      H      C   OH     H      C   OH      HO  C   H
H   C   OH      H      C   OH     H      C   OH      H   C   OH
    CH₂OH              CH₂OH              CH₂OH              CH₂OH

glucose           fructose           mannose            galactose
```

common hexose sugars

epimers (*n.pl.*) compounds with identical atoms but different configurations about one carbon atom, e.g. galactose and glucose.

glucose (*n*) a hexose (↑) sugar. In plants it is a product of photosynthesis and is stored as starch. Animals obtain glucose by digestion of carbohydrates; they store it as glycogen. Biological energy is obtained by the oxidation of glucose into carbon dioxide and water. Glucose is a component of disaccharides (e.g. sucrose) and polysaccharides (e.g. starch, cellulose, glycogen).

dextrose (*n*) one of the optical isomers (↑) of glucose.

fructose (*n*) a hexose (↑) sugar. It is a component of sucrose (↓). It is found widely in plants, particularly in fruits.

galactose (*n*) a hexose (↑) sugar. It is a component of lactose (↓) and is also found in plant polysaccharides, e.g. pectins. It is an epimer (↑) of glucose.

mannose (*n*) a hexose (↑) sugar. An epimer (↑) of glucose. It is a component of many polysaccharides.

sucrose, cane sugar (*n*) a non-reducing disaccharide composed of a glucose and a fructose molecule. It is widely found in plants and is their principal transport carbohydrate; it is not synthesised by animals.

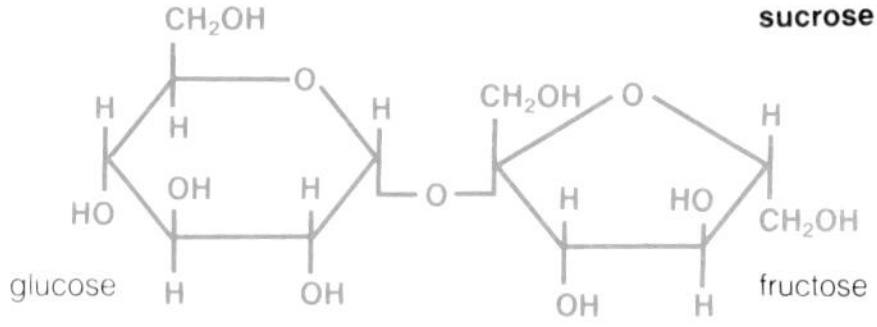

maltose (*n*) a reducing disaccharide composed of two chemically combined molecules of glucose. It is found in some germinating seeds and in the alimentary canal as a result of the action of maltase on starch.

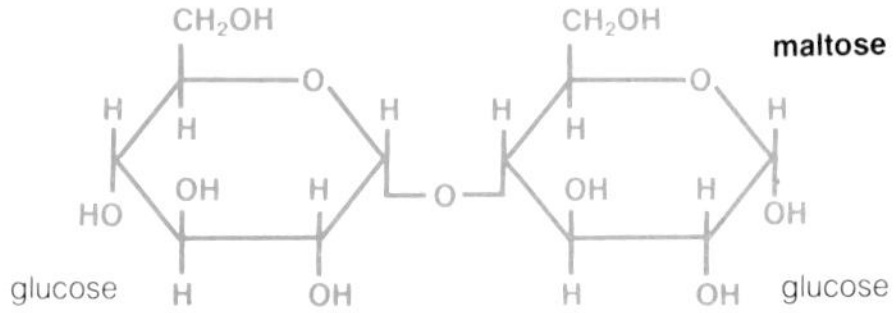

lactose, milk sugar (*n*) a reducing disaccharide composed of a glucose and a galactose molecule chemically combined.

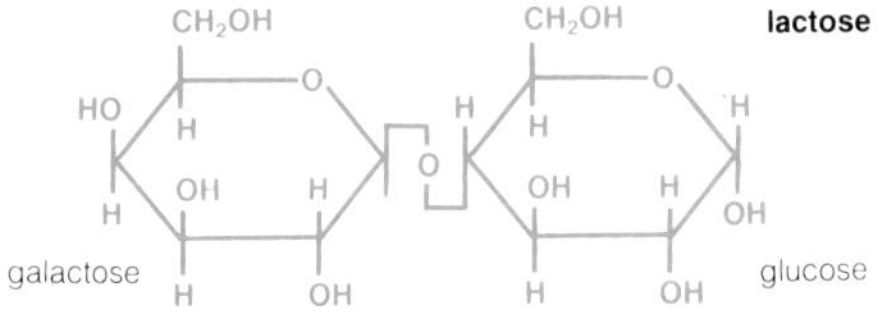

glucosamine (*n*) a hexose (↑) sugar containing an amino group, it is found in heparin and other polysaccharides.

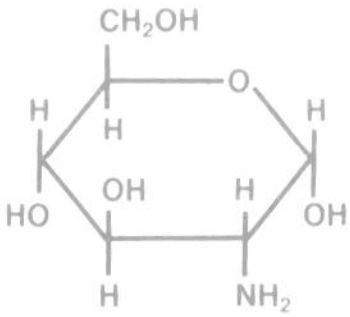

ribose (*n*) a pentose sugar; formula $C_5H_{10}O_5$ (*see diagram, p.115*). It is found in RNA (p.118).

nucleotide (n) an organic compound consisting of a pentose sugar (ribose or deoxyribose), phosphoric acid and a nitrogenous base (cytosine, thymine, uracil, adenine or guanine). The nucleic acids (↓) DNA and RNA (↓) consist of a chain of many nucleotides.

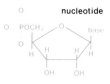

nucleotide

nucleoprotein (n) a large molecule consisting of a nucleic acid (↓), i.e. DNA or RNA (↓) bound to a protein.

nucleic acid (n) a long chain of many nucleotides (↑) chemically bound together. They are usually found as nucleoproteins (↑). There are two forms, DNA and RNA (↓).

∝ **helix, alpha helix** (n) a right-handed helix (↓). It is the configuration of many polypeptide chains and fibrous proteins. It is stabilized by cross links consisting mainly of hydrogen bonds.

helix (n) a series of concentric rings of the same diameter connected to form a cylinder. Compare spiral (↓).

helix

spiral (n) a series of concentric rings of different diameters connected together. The resulting structure may be flat or conical. Compare helix (↑).

spiral

RNA, ribonucleic acid (n) a nucleic acid (↑) consisting of a long chain of covalently-linked nucleotides. Each nucleotide consists of ribose (a sugar), phosphoric acid and one of four bases, adenine, guanine, cytosine or uracil. Uracil is the only base found in RNA but not DNA. RNA is made in the cell nucleus and is involved in protein synthesis (↓). There are several types of RNA including rRNA (↓), mRNA (↓) and tRNA (↓). In some viruses RNA is the genetic material instead of DNA.

uracil (n) a nitrogenous base found in RNA (↑).

rRNA, ribosomal RNA (n) type of RNA (↑) found in ribosomes (↓) which together with ribosomal proteins forms the structure of the ribosomes.

mRNA, messenger RNA (n) type of RNA (↑) which transfers genetic information from the nucleus to the ribosomes. The order of the bases of the DNA on which it is made determines the order of its bases. This in turn determines the order in which the amino acids are assembled in the protein being made.

tRNA, transfer RNA (n) type of RNA (↑) found in the cell cytoplasm which brings to the ribosome the amino acid molecules needed for protein synthesis (↓).

ribosome (n) tiny particle or organelle in cytoplasm of most living cells. Protein synthesis (↓) occurs on it. It consists of about equal amounts of RNA (↑) and protein.

RNA

segment of RNA chain

uracil

protein synthesis (*n*) the manufacture of protein molecules from their constituent amino acids (p.121). The information which determines the sequence of amino acids in a protein is stored in the DNA of the nucleus. The information is passed in the form of mRNA (↑) to the ribosomes in the cytoplasm. Proteins are made on the ribosomes.

protein (*n*) a chain of amino acids (p.121) joined together by **peptide bonds** to form a complex molecule. In fibrous protein, the chain is mostly in the form of an ∝helix (↑) held by hydrogen bonds (p.239). In globular proteins, the chain is much folded. In any one protein the sequence of amino acids is always the same; different arrangements of amino acids make different proteins. The sequence is genetically determined, *see protein synthesis* (↑). Proteins are present in all living cells, usually both as structural proteins, e.g. actomyosin and as enzymes, e.g. lipase, pepsin. Simple proteins consist only of amino acids, e.g. albumin, globulin. Proteins may, however, be combined with other substances, e.g. nucleic acid (nucleoproteins) carbohydrates (glycoproteins (p.120)), fats (lipoproteins), phosphoric acid (phosphoproteins (p.120)). These are the **conjugated proteins**. The presence of a protein is shown using Biuret, Millon's, xanthoproteic or ninhydrin (p.120) reactions. These are tests for specific amino acids, thus a negative result with a particular test may be obtained if the protein does not contain that particular amino acid. Thus one positive test identifies a protein even if the other tests are negative.

peptide (*n*) a compound of two or more amino acids (p.121) joined by a peptide bond. *Compare polypeptide* (↓).

polypeptide (*n*) strictly, a compound formed from three or more amino acids (p.121) joined together by peptide bonds, but in practice usually much more complex than a peptide (↑).

peptone (*n*) the product formed when proteins (↑) are only partly broken down by enzyme activity. Breakdown of peptones yields amino acids.

Biuret test (*n*) a biochemical test for protein (↑). An alkaline solution of copper (II) sulphate gives a purple colour with protein. In the absence of protein the solution remains blue.

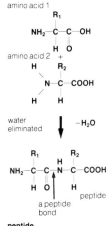

peptide
peptide linkage of amino acids

xanthoproteic test (*n*) biochemical test for protein
(p.119). Concentrated nitric acid added to protein gives
a yellow colour. When cooled, addition of ammonium
hydroxide changes this to bright orange.

Millon's test (*n*) a biochemical test for protein (p.119).
Protein when heated with Millon's reagent (mercuric
nitrate and nitrous acid) gives a brick red precipitate. In
the absence of protein the solution remains clear.

ninhydrin test (*n*) a biochemical and chromatographic
test for protein (p.119). Ninhydrin when heated with a
protein or amino acid produces a blue colour.

denaturation (*n*) structural changes occurring in a
protein (p.119) when it is subjected to heat, pH
changes or certain chemicals. Denaturation may lead
to a loss in biological activity and changes in physical
properties. **denature** (*v*).

denaturation

albumin (*n*) one of a group of simple proteins (p.119).
They are soluble in water and coagulated by heat. They
are found in egg white, i.e. albumen (↓), blood and milk.

albumen (*n*) egg white of birds and some reptiles. It is a
solution of several proteins (including albumin (↑)) in
water; found between yolk and membranes of shell.

globulin (*n*) one of a group of proteins (p.119); insoluble
in water but soluble in dilute salt solutions, coagulated
by heat, e.g. serum globulin.

glutelin (*n*) one of a group of simple proteins (p.119)
found in plants, e.g. gluten of wheat.

histone (*n*) one of a group of proteins (p.119) that
contain a large amount of basic amino acids; often
found in association with nucleic acids. They are
soluble in water and not easily coagulated by heat.

collagen (*n*) a fibrous protein (p.119). It forms the white
fibres in connective tissue. It has high tensile strength
but unlike elastin (↓) it is relatively inelastic. When boiled
it forms gelatin.

protein denaturation;
H bonds are broken
and the molecule
looses shape becoming
a chain

elastin (*n*) an elastic fibrous protein (p.119). It forms the
yellow fibres in connective tissue. It is found in elastic
structures, e.g. walls of large blood vessels, alveoli of
lungs. Unlike collagen (↑) it is resistant to boiling.

glycoprotein (*n*) a complex protein (p.119) containing
sugar molecules, e.g. some enzymes and hormones.

phosphoprotein (*n*) a complex protein (p.119)
containing phosphoric acid, e.g. casein (↓).

casein (*n*) the main protein (p.119) of milk. It is a
phosphoprotein (↑) which is precipitated by rennin.

amino acid
general formula
for an amino acid

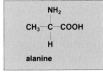

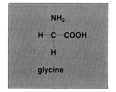

amino acid (*n*) organic acid with a free, acidic carboxyl group ($-COOH$) and a free, basic amino group ($-NH_2$). The R-side group may be: a hydrocarbon (e.g. alanine), a hydroxyl (e.g. serine), aromatic (e.g. phenyl-alanine), acidic (e.g. aspartic acid), basic (e.g. arginine), sulphur-containing (e.g. methionine). They are important fundamental constituents of living matter because they are combined together to make proteins. There are many naturally occurring amino acids but only twenty (see below) are commonly found in proteins. Some, the essential amino acids (↓), must be obtained from the diet because the organism cannot synthesise them. The others, the non-essential amino acids, can be synthesised by the organism.

essential amino acid (*n*) amino acid which cannot be synthesised by animals and must be supplied in the diet. For man they are isoleucine, leucine, lysine, methionine, phenylalanine, threonine, tryptophan and valine. The list is almost the same for other animals.

alanine (*n*) a neutral amino acid (↑); one of the twenty common ones found in proteins.

arginine (*n*) a basic amino acid (↑); one of the twenty common ones found in proteins.

asparagine (*n*) a basic amino acid (↑); one of the twenty common ones found in proteins.

aspartic acid (*n*) an acidic amino acid (↑); one of the twenty common ones found in proteins. It also serves as a precursor for pyrimidine nucleotide synthesis.

cysteine (*n*) a sulphur-containing amino acid; one of the twenty common ones found in proteins. It is often at the active site of enzymes. Two adjacent cysteine molecules joined by a disulphide bridge form **cystine**; this links polypeptides together to form complex proteins.

glutamic acid (*n*) an acidic amino acid (↑); one of the twenty common ones found in proteins. It is a precursor in the synthesis of other amino acids.

glutamine (*n*) an amino acid (↑); one of the twenty common ones found in proteins.

glycine (*n*) the simplest amino acid (↑); one of the twenty common ones found in proteins.

histidine (*n*) a basic amino acid (↑); one of the twenty common ones found in proteins.

isoleucine (*n*) a neutral amino acid (↑); one of the twenty common ones found in proteins. It is an essential amino acid (↑) for man.

leucine (*n*) a neutral amino acid (p.121); one of the twenty common ones found in proteins. It is an essential amino acid (p.121) for man.

lysine (*n*) a basic amino acid (p.121); one of the twenty common ones found in proteins. It is an essential amino acid (p.121) for man.

methionine (*n*) a sulphur-containing amino acid (p.121); one of the twenty common ones found in proteins. It is an essential amino acid for man.

ornithine (*n*) a basic amino acid (p.121); not found in proteins.

phenylalanine (*n*) an aromatic amino acid (p.121); one of the twenty common ones found in proteins. It is the immediate precursor of tyrosine (↓) and is an essential amino acid (p.121) for man.

proline (*n*) an amino acid (p.121); one of the twenty common ones found in proteins. It has a cyclic structure.

serine (*n*) a hydroxyl amino acid (p.121); one of the twenty common ones found in proteins.

threonine (*n*) a hydroxyl amino acid (p.121); one of the twenty common ones found in proteins. It is an essential amino acid (p.121) for man.

tryptophan (*n*) an aromatic amino acid (p.121); one of the twenty common ones found in proteins. It is an essential amino acid (p.121) for man.

tyrosine (*n*) a hydroxyl, aromatic amino acid (p.121); one of the twenty common ones found in proteins. It is used in the synthesis of adrenaline, thyroid hormone and alkaloids like morphine.

valine (*n*) a neutral amino acid (p.121); one of the twenty common ones found in proteins. It is an essential amino acid (p.121) for man.

micronutrient (*n*) a nutrient, required only in small amounts, that is necessary for the healthy growth of an organism, e.g. trace elements (↓) and vitamins.

trace element (*n*) an element needed in very small (i.e. **trace**) amounts for growth of an organism. Trace elements include iron, copper, zinc, molybdenum, manganese, boron; most of these are essential for both plants and animals. Plants obtain trace elements from the soil, animals from their diet. In plants and animals they are cofactors (p.109) of enzymes, in animals they are also components of hormones. Lack of trace elements results in disease and even death.

methionine

phenylalanine

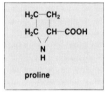

proline

mineral salts (*n.pl.*) inorganic salts needed by plants and animals for healthy functioning of the organism. Plants obtain them from the soil, animals from their diet. Included are ions of the elements calcium, sodium, potassium, phosphorus, chlorine, iodine, magnesium, fluorine, sulphur. Lack of mineral salts results in disease; e.g. in man lack of calcium and phosphorus causes rickets, lack of calcium prevents blood clotting, lack of iron causes anaemia.

mineral (*n*) a naturally occurring substance produced by inorganic processes; it is not of animal or plant origin.

roughage (*n*) fibre, bran, etc. necessary to promote intestinal movement. It gives bulk to the faeces. Usually it is provided in the diet in the form of cellulose, e.g. in wholemeal cereals and vegetables.

lipid (*n*) one of a large group of organic substances found in plants and animals. They are soluble in fat solvents (ether, hot ethanol, petrol); insoluble in water. Included are simpler straight chain molecules of fats (↓) and waxes (↓) and the more complex phospholipids (p.124) and steroids (↓).

fat (*n*) (1) an organic substance which is extracted from plant or animal tissues by fat solvents (ether, hot ethanol, petrol). It consists of fatty acids (p.124) and glycerol (p.125) and contains the elements carbon, hydrogen and oxygen. True fats are solid below 20°C; when liquid below 20°C they are called **oils**. Oils have a higher proportion of unsaturated (p.125) fatty acids than solid fats. (2) an alternative term for adipose tissue which is full of fats and acts as an insulating layer and energy store in higher animals and some plants. *See lipid* (↑).

wax (*n*) one of a group of organic compounds produced by plants and animals. They contain the elements carbon, hydrogen and oxygen and are esters of fatty acids (p.124) with long chain alcohols. They often act as a protective, outer covering, e.g. beeswax. Waxes in plants form water-repellent layers on leaves, fruits and seeds. *See lipid* (↑).

steroid (*n*) one of a group of complex organic compounds found in plants and animals. They are hydrocarbons in which the carbon atoms are arranged in a system of rings. Steroids include vitamin D, hormones and bile salts. The steroid alcohols are called **sterols**, e.g. cholesterol, cortisone (p.124).

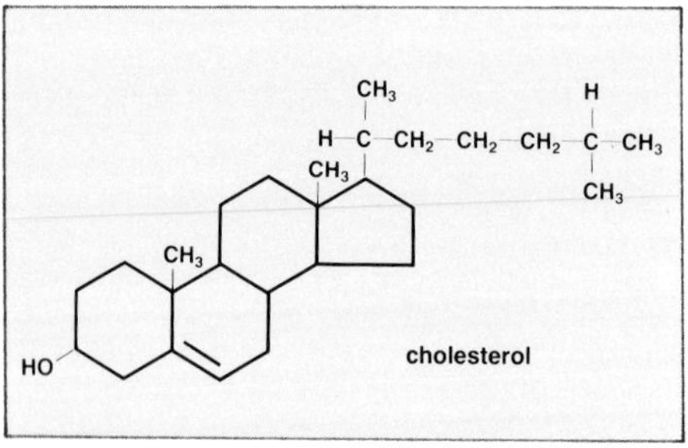

CH₃

H—C—CH₂—CH₂—CH₂—C—CH₃

cholesterol

cholesterol (*n*) an animal sterol (p.123). It is found in plasma membranes, bile, blood cells, egg yolk.

cortisone (*n*) a steroid (p.123) hormone made by the cortex of the adrenal gland. It functions in cell metabolism, diminishes local inflammation and helps healing of wounds. It is used to treat rheumatoid arthritis.

ergosterol (*n*) a plant sterol (p.123).

phospholipid (*n*) a complex, organic lipid (p.123) molecule, e.g. lecithin. Phospholipids contain glycerol (↓), fatty acids (↓), phosphoric acid and a nitrogenous base. They are commonly found in cell membranes.

choline (*n*) a basic compound found in phospholipids (↑) like lecithin (↓) and in acetycholine (p.165).

lecithin (*n*) a phospholipid (↑). It contains glycerol (↓), fatty acid (↓), choline (↑) and phosphoric acid. It is found in all animal and plant cells.

lipoprotein (*n*) a substance consisting of protein combined with a lipid (p.123) found, e.g. in blood, lymph and membranes.

glycolipid (*n*) a lipid (p.123) that contains one or more sugar groups.

fatty acid (*n*) a carboxylic acid. Biologically fatty acids usually contain an even number of carbon atoms arranged in a straight chain. There are three types. (1) **saturated fatty acids**, e.g. palmitic acid (↓), stearic acid (↓) with no double bonds between carbon atoms. (2) **unsaturated fatty acids**, e.g. oleic acid (↓), linoleic acid (↓) with one or two double bonds linking carbon atoms. (3) **polyunsaturated fatty acids** with more than two double bonds. Long chain fatty acids are insoluble in water but their sodium and potassium salts, the so-called **soaps**, are soluble.

saponification (n) hydrolysis of glycerides ($\downarrow$) by alkalis to give glycerol ($\downarrow$) and a salt of a fatty acid ($\uparrow$). The latter is a soap.

saponification

$$H_2C\text{—}O\text{—}CO\text{—}R \qquad\qquad\qquad H_2C\text{—}OH$$
$$HC\text{—}O\text{—}CO\text{—}R\ +\ 3KOH \rightarrow 3RCOOK\ +\ HC\text{—}OH$$
$$H_2C\text{—}O\text{—}CO\text{—}R \qquad\qquad\qquad H_2C\text{—}OH$$

glyceride   alkali   a soap   glycerol

R varies according to the triglyceride used

saturated (adj) (1) describes a molecule where all valence electrons of the carbon atoms are paired, i.e. there are no double or triple bonds. (2) describes a solution containing the maximum amount of solute at a given temperature. *Contrast unsaturated* ($\downarrow$).

unsaturated (adj) (1) describes a molecule which contains double or triple bonds between carbon atoms. (2) describes a solution in which more solute can be dissolved at that temperature. *Contrast saturated* ($\uparrow$).

palmitic acid (n) a saturated fatty acid ($\uparrow$). Formula $CH_3(CH_2)_{14}COOH$.

stearic acid (n) a saturated fatty acid ($\uparrow$). Formula $CH_3(CH_2)_{16}COOH$.

linoleic acid (n) an unsaturated fatty acid ($\uparrow$). Formula $CH_3(CH_2)_4-CH = CHCH_2CH = CH(CH_2)_7COOH$. It is needed in very small amounts, probably by all animals. It is found in many vegetable oils. See vitamin F (p.128).

oleic acid (n) an unsaturated fatty acid ($\uparrow$). Formula $CH_3(CH_2)_7CH = CH(CH_2)_7COOH$.

glycerol (n) a sugar alcohol. Formula: *see diagram*. One, two or all three of its hydroxyl groups can be substituted with organic acids to form esters. In fats (p.123) such as triglycerides all three hydroxyls are substituted with fatty acids ($\uparrow$).

glyceride (n) a fatty ester of glycerol ($\uparrow$). One (monoglyceride), two (diglyceride) or all three (triglyceride) of the hydroxyl groups of glycerol may be substituted.

Sudan III (n) a red dye which is taken up by fats and is therefore used as a test for them.

grease spot test (n) test for fats and oils. When fats and oils are smeared on paper they leave a permanent translucent patch.

$$CH_2OH$$
$$|$$
$$CHOH$$
$$|$$
$$CH_2OH$$

glycerol

vitamin (*n*) a complex, organic micronutrient (p.122)
required in small amounts by many heterotrophs, e.g.
animals, fungi, bacteria. Each organism has its own
specific requirements for vitamins. Occasionally some
organisms can synthesise a particular vitamin, e.g. man
can synthesise vitamin D. Where this is not possible
they must be obtained from the environment or in the
diet. Vitamins are essential for metabolic functions,
often acting as a cofactor of enzyme reactions. They
are divided into water soluble vitamins (i.e. the B
complex vitamins and ascorbic acid) and fat soluble
vitamins (i.e. vitamins A, D, E, K). An insufficient amount
of vitamins in the diet of animals leads to vitamin
deficiency diseases like beri-beri, scurvy, rickets,
pellagra *(see p.128, 129)*.

vitamin A, retinol (*n*) a fat soluble vitamin (↑) required by
man and other vertebrates but not other animals. The
major source of vitamin A is the carotene pigments
found in plants, e.g. in carrots. Animals store it in the
liver; it is especially abundant in fish livers. Mammals
use it to make visual purple, a pigment needed for
vision. Deficiency in the diet results in night blindness,
xerophthalmia (p.128).

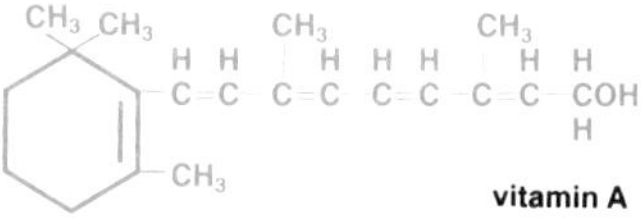

vitamin A

vitamin B complex (*n*) a group of water soluble vitamins
(↑) that were first thought to be a single compound.
Included are vitamins B_1, B_2, B_6, B_{12}, nicotinic acid,
pantothenic acid, biotin, folic acid.

vitamin B_1, thiamine (*n*) vitamin of B complex (↑),
required by man, vertebrates and insects. It is found in
yeast, bran from cereals and pulses. It acts as a
coenzyme in breakdown of carbohydrate. Deficiency in
man causes beri-beri (p.128).

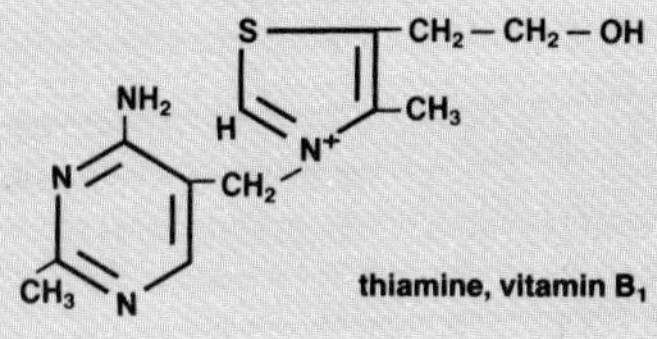

thiamine, vitamin B_1

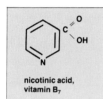

pyridoxine, vitamin B₆

nicotinic acid,
vitamin B₇

vitamin B₂, riboflavin (*n*) vitamin of B complex (↑) required by man, vertebrates, insects and some bacteria. It is found in yeast, milk, liver, green vegetables. Deficiency causes conjunctivitis.

vitamin B₆, pyridoxine (*n*) vitamin of B complex (↑), required by mammals, birds, insects, bacteria and some yeasts. It is found in yeast, rice polishings and egg yolk. It forms a coenzyme involved in decarboxylation and transamination of amino acids.

vitamin B₇, nicotinic acid, niacin (*n*) vitamin of B complex (↑), required by man,.vertebrates and insects. It is found in milk, yeast, liver, whole cereals. It is used in synthesis of NAD and NADP. Deficiency results in pellagra (p.128).

vitamin B₁₂, cobalamin (*n*) vitamin of B complex (↑) required by many animals. It contains cobalt. It is found in liver, lean meat, fish, milk; plants do not contain it. Deficiency affects cell division and red blood cell formation (see pernicious anaemia p.145).

pantothenic acid (*n*) vitamin of B complex (↑) required by man, vertebrates, insects, some bacteria, some yeasts. It is abundant in yeast, liver, eggs. It acts as a coenzyme in glycolysis.

co A, coenzyme A (*n*) a coenzyme synthesised from pantothenic acid (↑). Has role in Krebs cycle and in fatty acid synthesis and oxidation.

biotin (*n*) vitamin of B complex (↑). It is abundant in egg yolk, liver, yeast. Bacteria in the large intestine of man can produce it. It acts as a coenzyme in reactions where carboxyl groups are removed or added.

folic acid

folic acid (*n*) vitamin of B complex (↑). It is found in green, leafy vegetables and liver. It is reduced to form the active coenzyme concerned in methylation reactions. Deficiency causes a type of anaemia.

vitamin C, ascorbic acid (*n*) a water soluble vitamin (↑), required by man, primates and guinea pigs; other animals can synthesise it. It is found in fresh, green vegetables and citrus fruit. Deficiency causes scurvy (p.129), weakness of capillary walls, lowered resistance to disease and poor healing of wounds.

ascorbic acid, vitamin C

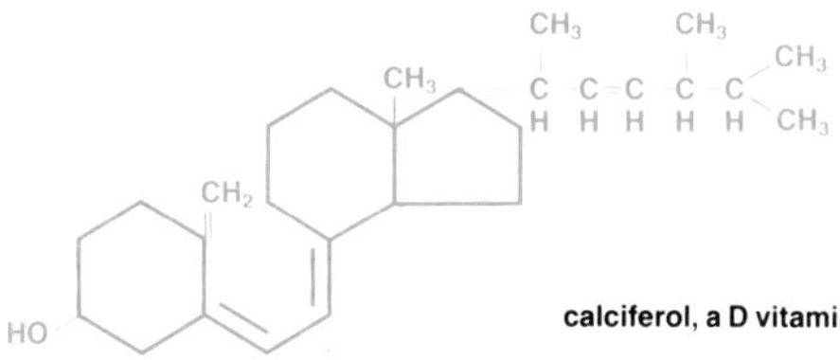

calciferol, a D vitamin

vitamin D (*n*) fat soluble vitamin (p.126) consisting of several related steroids including **calciferol**. It is found in fish liver, eggs, liver, butter, cheese. It is required by man (although some is made by the action of ultraviolet radiation on the skin) and all other vertebrates for absorption and assimilation of calcium and phosphate ions from food. Deficiency causes rickets (↓) and osteomalacia (↓).

vitamin E (*n*) fat soluble vitamin (p.126) required by vertebrates. It is found in wheat germ and some vegetable leaves. Deficiency causes abortion and sterility in rats.

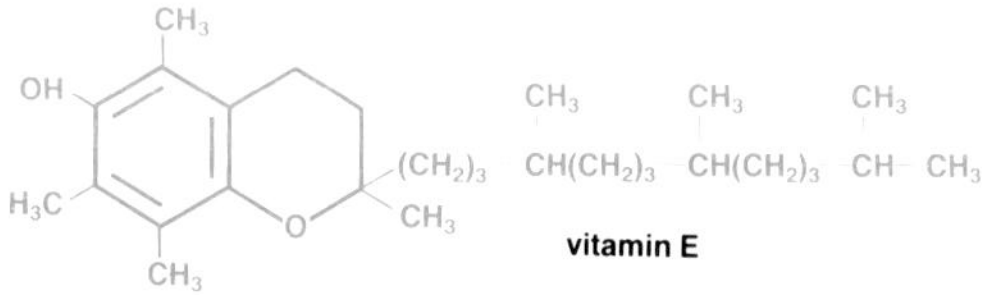

vitamin E

vitamin F (*n*) alternative term for linoleic acid (p.125).
vitamin K (*n*) fat soluble vitamin (p.126) required by man, other mammals and birds; found in all dark-green, leafy vegetables. Bacteria in human gut can synthesise part of man's requirement. It takes part in synthesis of prothrombin in the liver; deficiency leads to defective blood clotting.

night blindness (*n*) deficiency disease (p.214) in man caused by lack of vitamin A (p.126) in the diet. Vision is impaired at night due to deficiency of visual purple in the rods (p.173).

xerophthalmia (*n*) deficiency disease (p.214) in man due to lack of vitamin A (p.126) in the diet. It affects the cornea (p.171) and may cause blindness.

beri-beri (*n*) deficiency disease (p.214) in man due to lack of vitamin B_1 (p.126) in the diet.

pellagra (*n*) deficiency disease (p.214) in man due to lack of vitamin B_7 (p.127) or tryptophan in the diet.

scurvy (*n*) deficiency disease (p.214) in man due to lack of vitamin C (p.127) in the diet. It causes loosening of teeth and bleeding under the skin, especially in gums.

rickets (*n*) deficiency disease (p.214) in young children. It is caused by failure to absorb calcium as a result either of lack of vitamin D or calcium in the diet. Bones become soft and limbs bend under the child's weight.

osteomalacia (*n*) deficiency disease (p.214) in adults, especially pregnant women. It is caused by failure to absorb calcium as a result either of lack of vitamin D or calcium in the diet. Bones become soft.

goitre (*n*) deficiency disease (p.214) in man due to lack of iodine in the diet or inability of the thyroid gland to secrete sufficient thyroxine. Causes enlarged thyroid.

malnutrition (*n*) faulty nutrition (↓) of animals in which either an unbalanced diet (↓) provides too much or too little nutrients (↓) for the body or the body does not assimilate the digested food from a balanced diet.

undernourishment (*n*) lack of a sufficient supply in the diet of one or more nutrients (↓) needed for the maintenance of health and growth of an organism.

nutritional requirements, nutriment (*n*) all the essential nutrients (↓) needed by an organism for nutrition (↓).

nutrient (*n*) a food material; part of the nutritional requirement of an organism, e.g. mineral salts are nutrients for plants; glucose, vitamins, etc., are nutrients for animals.

nutrition (*n*) process by which an organism obtains nutrients (↑) it requires to carry out its vital functions.

diet (*n*) the food type, quality and quantity that is eaten by an animal. **Balanced diets** supply the animal with optimum amounts of the different types of nutrient (↑) it requires. In man, these requirements vary with factors such as age, sex and occupation. A diet which supplies insufficient or excess of a particular nutrient(s) is said to be an **unbalanced diet**.

appetite (*n*) the desire, especially in humans, for food. Disease often results in poor appetite.

calorie (*n*) the amount of heat required to raise a gram of water from 15°C to 16°C. Formerly used as a measure of heat or energy value of foods. It is now replaced by the joule (↓). 1000 calories = 1 kilocalorie = 1C.

joule (*n*) a unit of energy, work or heat. It has replaced the calorie (↑) as a measure of the heat or energy value of food and energy turnover in animals. 4.2 joules = 1 calorie; 1000 joules = 1 kilojoule.

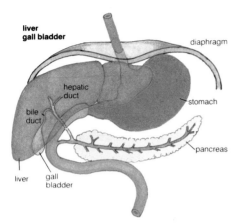

liver (*n*) a large gland in animals that opens into the gut. It has many functions in the organism including: bile production; deamination ($\downarrow$) of amino acids with the formation of urea ($\downarrow$); carbohydrate storage in the form of glycogen; storage of minerals and vitamins; detoxication of foreign chemicals; synthesis of fibrinogen and prothrombin; breakdown of haemoglobin. These metabolic activities are the main source of the heat which maintains the body temperature in mammals.

hepatic (*adj*) concerned with the liver, e.g. the **hepatic duct**, which leaves the liver and conducts bile to the gall bladder ($\downarrow$).

gall bladder (*n*) a small gland found in many vertebrates between the lobes of the liver ($\uparrow$). The liver secretes bile continuously; this passes via ducts to the gall bladder where it is stored. The contractile walls of the gall bladder expel the bile into the intestine when food, especially fat, is present in the intestine.

$$R - \underset{\underset{NH_2}{\displaystyle|}}{\overset{\overset{H}{\displaystyle|}}{C}} - COOH + \tfrac{1}{2}O_2 \rightarrow R - CO - COOH + NH_3$$

R depends on the amino acid
deamination of amino acids   **deamination**

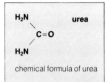

$$H_2N$$
$$\backslash$$
$$C=O \quad \text{urea}$$
$$/$$
$$H_2N$$

chemical formula of urea

deamination (*n*) the process of removal of an amino group ($-NH_2$) from amines such as amino acids. In mammals the deaminating enzymes which do this are mainly in the liver (↑); the ammonia liberated is converted by it into urea (↓). Other vertebrates, e.g. birds, convert it to uric acid (↓).

urea (*n*) a water soluble, organic compound. It is produced from the ammonia formed in the liver by deamination (↑) of amino acids and thus represents the main nitrogenous waste product of protein breakdown in some vertebrates, e.g. mammals, some fish. It passes from the liver into the blood and is extracted from this by the kidney and excreted in the form of urine. *Compare uric acid* (↓).

uric acid (*n*) an organic compound, slightly soluble in water. It is produced from the ammonia formed in the liver by deamination (↑) of amino acids and thus is the main excretory product of protein and nucleic acid breakdown in terrestrial animals that develop in shells and so cannot get rid of aqueous excretory products as urea (↑) or ammonia, e.g. birds, insects. In primates (e.g. man) it is the end-product of nucleic acid breakdown.

ureotelic (*adj*) describes animals that eliminate nitrogenous waste as urea (↑), e.g. man.

uricotelic (*adj*) describes animals that eliminate nitrogenous waste as uric acid (↑), e.g. birds.

ammoniotelic (*adj*) describes animals that eliminate nitrogenous waste as ammonia, e.g. protozoans, many invertebrates.

urease (*n*) enzyme found in many plants and a few invertebrate animals that splits urea into ammonia and water.

ornithine cycle (*n*) cycle of reactions in vertebrate liver which converts ammonia into urea (↑).

detoxication (*n*) chemical modification of a toxic substance into a less toxic product. In vertebrates it is usually carried out in the liver.

jaundice (*n*) disease in man caused by malfunction of the liver. Excess bile pigments in blood and tissues result in a yellow colouration to the skin and eyes.

hepatitis (*n*) inflammation of the liver in mammals; often the result of infection by a virus (**viral hepatitis**).

gall stone (*n*) hard stone formed in the gall bladder (↑). It contains cholesterol.

vascular system (*n*) in vertebrates a system of vessels to conduct liquids. The blood system carries blood, the lymphatic system transports lymph.

circulatory system (*n*) the blood vascular system (↑) of animals, especially vertebrates. It is a closed system made up of arteries (↓), arterioles (↓), capillaries (↓), venules (↓) and veins (↓).

blood vessels (*n*) tubes which carry blood in vertebrates, e.g. arteries (↓), arterioles (↓), capillaries (↓), venules (↓), veins (↓).

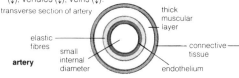

transverse section of artery

elastic fibres

small internal diameter

artery

thick muscular layer

connective tissue

endothelium

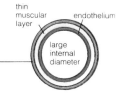

vein

transverse section of vein

thin muscular layer

endothelium

large internal diameter

artery (*n*) a blood vessel (↑) which carries blood from the heart. Large arteries leave the heart, branch into smaller arteries then further divide into arterioles (↓). Vertebrate arteries have thick, elastic muscular walls to withstand high blood pressure; they are lined with endothelium (p.16).

arteriole (*n*) a small artery (↑). Arterioles are continuous with the capillaries (↓).

capillary, blood capillary (*n*) very small blood vessel (↑) that receives blood from arterioles (↑) and passes it to venules (↓). Capillaries are numerous, forming an inter-communicating network in almost all vertebrate tissues. Their walls consist of a single layer of endothelium (p.16). The main exchange of water, glucose, amino acids, inorganic ions and dissolved gases (oxygen, carbon dioxide) between blood and tissues occurs across them.

venule (*n*) a small vein (↓). Venules collect blood from capillaries (↑); they unite to form veins (↓).

vein (*n*) a blood vessel (↑) which carries blood from venules (↑) to the heart. Veins are lined with endothelium (p.16); walls are thinner and internal diameters are larger than in arteries. Unlike arteries and capillaries, valves are present to prevent backflow of blood and ensure that it flows only towards the heart.

elastic (*adj*) describes an object, e.g. artery wall, that after stretching can recover its original size and shape.

elasticity (*n*) ability to return to original form and size after stretching.

vein

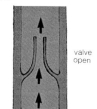

valve open

action of valves in veins

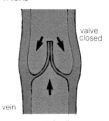

valve closed

vein

← direction of blood flow

vasoconstriction (*n*) a decrease in diameter of blood vessels (↑), particularly arterioles (↑), brought about by contraction of smooth muscle in their walls. It is caused by stimulation of vasomotor nerves (↓), the hormone adrenalin, or other factors, e.g. cold, pain, fall in blood pressure. *Contrast vasodilation* (↓).

vasodilation (*n*) an increase in diameter of blood vessels (↑), particularly arterioles (↑), brought about by relaxation of smooth muscle in their walls. It is caused by an increase in body temperature, rise in blood pressure or other factors. *Compare vasoconstriction* (↑).

vasomotor nerves (*n*) the nerves of the autonomic nervous system which when stimulated alter the diameter of blood vessels (↑).

blood (*n*) a fluid tissue transported in blood vessels (↑) of an animal. In vertebrates it is circulated by the action of the heart. Vertebrate blood is composed of liquid plasma containing red and white blood cells, platelets (in mammalian blood), dissolved respiratory gases, digestive and excretory products, proteins, hormones, mineral salts. Blood has three main functions: transport; regulation, e.g. it helps to control body temperature; defence against pathogens and poisons.

blood cell (*n*) a cell that circulates in blood plasma. It was formerly known as a **blood corpuscle**. There are two types, red and white blood cells.

platelet (*n*) a very small particle, found only in mammalian blood. Platelets are non-nucleated fragments of cells from red bone marrow. There are about 250000 per mm^3 of human blood. On disintegration they liberate thrombokinase, essential for blood clotting. This function in non-mammalian blood is performed by thrombocytes (p.140).

haemoglobin (*n*) a red respiratory pigment found in the red blood cells of vertebrates and in the plasma of some invertebrates. It consists of iron-containing haem (↓) combined with a blood protein, globin. It is related chemically to chlorophyll of plants, myoglobin of voluntary muscle and cytochromes. It transports oxygen; when deoxygenated, i.e. not containing oxygen, it is bluish red; when oxygenated it is scarlet and is known as **oxyhaemoglobin**.

haem, heme (*n*) an iron-containing organic molecule found in proteins such as haemoglobin (↑), myoglobin and the cytochromes.

red blood cell

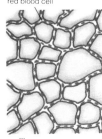

capillary
capillary network

heart

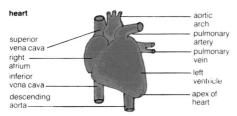

- aortic arch
- pulmonary artery
- pulmonary vein
- left ventricle
- apex of heart

superior vena cava
right atrium
inferior vena cava
descending aorta

heart
human heart, internal structure

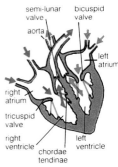

semi-lunar valve
bicuspid valve
aorta
left atrium
right atrium
tricuspid valve
right ventricle
chordae tendinae
left ventricle

→ deoxygenated blood flow
→ oxygenated blood flow

heart (*n*) a muscular pump of the vascular system (p.132). It possesses valves to maintain a one-directional blood flow. Insects have a long tubular heart. Different vertebrate classes have two, three or four chambered hearts. The human heart is shown in the diagram. Vertebrate hearts are composed of cardiac muscle (p.96) which contracts rhythmically, a phase of contraction, systole (↓), being followed by a phase of relaxation, diastole (↓). The contraction rate is controlled indirectly by the pacemaker (↓).

apex (*n*) the tip of a pointed structure, e.g. apex of heart.

pericardial cavity (*n*) body cavity in vertebrates containing the heart (↑).

pericardium (*n*) a membrane that forms the wall of the pericardial cavity (↑) in vertebrates.

atrium (*n*) one of the chambers in a vertebrate heart (↓); it receives blood from veins and pumps it into the ventricle (↓). The muscular wall is thinner and has a less powerful pumping action than that of the ventricle. Fish have one atrium; other vertebrates have two. Hearts of the latter receive oxygenated blood from the lungs in the left atrium and deoxygenated blood from the rest of the body in the right atrium.

auricle (*n*) alternative name for atrium (↑). Atrium is normally used in humans.

ventricle (*n*) (1) one of the chambers of the vertebrate heart (↑). It receives blood from the atrium (↑) and pumps it from the heart into the arteries. Its muscular wall is thicker and has a more powerful pumping action than that of the atrium. Fish and Amphibia have one ventricle; in most reptiles it is partially divided; birds and mammals have two. In mammals the right one pumps deoxygenated blood to the lungs, the left oxygenated blood to the rest of the body. (2) one of several cavities in a vertebrate brain. They are filled with cerebrospinal fluid.

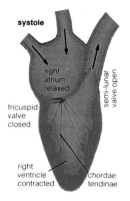

systole

right atrium relaxed

semi-lunar valve open

tricuspid valve closed

right ventricle contracted

chordae tendinae

deoxygenated blood flow

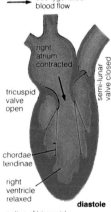

right atrium contracted

semi-lunar valve closed

tricuspid valve open

chordae tendinae

right ventricle relaxed

diastole

action of tricuspid and semi-lunar valves

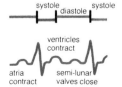

systole diastole systole

ventricles contract

atria contract

semi-lunar valves close

an ECG tracing

bicuspid valve, mitral valve (*n*) heart (↑) valve found between the left atrium (↑) and the left ventricle (↑) of mammals and birds. It has two membranous flaps which prevent blood flowing from ventricle to atrium when the ventricles contract.

tricuspid valve (*n*) heart (↑) valve found between the right atrium (↑) and right ventricle (↑) of mammals. It has three membranous flaps which prevent blood flowing from ventricle to atrium when the ventricles contract.

chordae tendinae (*n*) tendon-like cords that connect the flaps of the bicuspid (↑) and tricuspid (↑) valves to the walls of the ventricles (↑).

semi-lunar valves (*n.pl.*) valves found at the entrances to the pulmonary artery and aorta in vertebrates. They consist of three half-moon shaped membranous flaps attached to the artery wall. They prevent backflow of blood into the heart (↑).

pacemaker (*n*) in vertebrates a group of muscle cells in the heart (↑). It starts heartbeat (↓), causing contraction of the atria (↑). In mammals and birds it is in the wall of the right atrium near the vena cavae (p.136).

bundle of His (*n*) a bundle of nerve and muscle fibres in a vertebrate heart (↑). It carries impulses from the muscle of the atria (↑) to the muscle of the ventricles (↑) so that the ventricles contract after the atria.

heartbeat (*n*) in vertebrates the sound made when the heart (↑) pumps blood. The first sound of the double beat is caused by the ventricles contracting, the other by the semi-lunar valves (↑) closing. The **accelerator**, a nerve of the sympathetic nervous system, increases the rate of heartbeat; the **inhibitor**, a branch of the vagus nerve of the parasympathetic nervous system, decreases it.

systole (*n*) in vertebrates the phase of the heartbeat (↑) when the cardiac muscle (p.96) contracts and blood flows into the arteries. It is followed by diastole (↓).

diastole (*n*) in vertebrates the phase of the heartbeat (↑) when the cardiac muscle (p.96) relaxes and the heart fills with blood from the veins. It occurs after systole (↑).

electrocardiograph (*n*) an instrument that records, in the form of an **electrocardiogram** or an **ECG**, the electrical changes that occur when the heart beats.

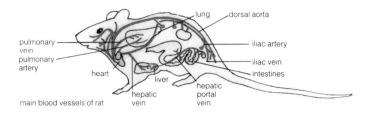

main blood vessels of rat

coronary vessels (*n*) vessels that supply the tissues of the vertebrate heart with blood.

cardiac (*adj*) concerning the heart, e.g. cardiac muscle (p.96).

cardiograph (*n*) an instrument that records the heartbeat (p.135). It produces a trace call a **cardiogram**.

aorta (*n*) the large artery in mammals, including man, that carries oxygenated blood from the left ventricle of the heart to all parts of the body, except the lungs. Many arteries branch from it.

dorsal aorta (*n*) an artery found in vertebrates. It passes oxygenated blood to the trunk and hind limbs. In man it is often called the descending aorta.

ventral aorta (*n*) a large artery in fish that carries deoxygenated blood from the ventricle. It gives off branches to the aortic arches (↓).

aortic arches (*n*) the six pairs of arteries in fish that connect the ventral aorta (↑) via the capillary network of the gills to the dorsal aorta (↑).

systemic arch (*n*) the fourth aortic arch (↑) in vertebrates, other than fish; in adults it supplies blood to all parts of the body except the head. In adult reptiles and Amphibia both right and left arches are present; birds have only the right one; mammals only the left, called the aorta (↑).

pulmonary artery (*n*) blood vessel in air-breathing vertebrates that carries deoxygenated blood from the heart to the lung capillaries. In crocodiles, birds and mammals it leaves the right ventricle.

pulmonary vein (*n*) blood vessel in air-breathing vertebrates that carries oxygenated blood from the lung to the left atrium of the heart.

vena cavae, caval veins (*n*) in air-breathing vertebrates the main veins collecting deoxygenated blood from all parts of the body and passing it to the right atrium of the heart. The single **posterior** or **inferior vena cava** collects blood from the hind limbs and trunk, the two (in some mammals, e.g. man, there is only one) **anterior** or **superior vena cavae** collect blood from the fore limbs and head.

hepatic portal system (*n*) system of veins in vertebrates that collects blood containing the absorbed products of digestion (except fats) from the small intestine. The veins unite to form the **hepatic portal vein** which leads to the liver (p.130). Liver cells perform many important functions on the dissolved substances, e.g. blood glucose is converted to glycogen. Liver cells also receive oxygenated blood from the **hepatic artery**. Blood leaves the liver in the **hepatic vein**.

portal vein (*n*) a vein that carries blood from one capillary network to another, e.g. the hepatic portal vein (↑), the renal portal vein (↓).

renal portal system (*n*) in fish, reptiles and amphibians a system of veins that take blood from the hind limbs and tail. They unite to form the **renal portal veins** which supply blood to the kidneys.

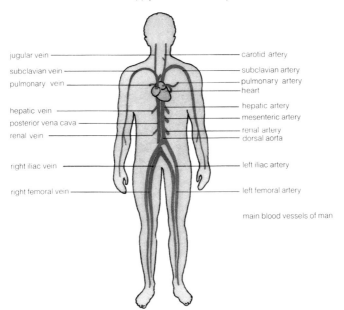

jugular vein
subclavian vein
pulmonary vein
hepatic vein
posterior vena cava
renal vein
right iliac vein
right femoral vein

carotid artery
subclavian artery
pulmonary artery
heart
hepatic artery
mesenteric artery
renal artery
dorsal aorta
left iliac artery
left femoral artery

main blood vessels of man

innominate artery (*n*) a short blood vessel that branches from the aorta of many birds and mammals. It divides into the subclavian (↓) and carotid (↓) arteries.

carotid artery (*n*) a vertebrate blood vessel supplying oxygenated blood to the head. In mammals there is one on either side of the neck.

subclavian artery (*n*) a blood vessel in vertebrates that supplies oxygenated blood to the fore limbs or pectoral fins of a fish.

mesenteric artery (*n*) a vertebrate blood vessel that branches from the aorta and supplies oxygenated blood to the intestines.

renal artery (*n*) a vertebrate blood vessel that branches from the aorta and supplies oxygenated blood to the kidney.

iliac artery (*n*) a blood vessel in air-breathing vertebrates that supplies oxygenated blood to the hind limbs. In mammals it becomes the **femoral artery** of the leg.

subclavian vein (*n*) a blood vessel in air-breathing vertebrates that carries deoxygenated blood from the fore limbs towards the heart.

jugular vein (*n*) a vertebrate blood vessel that carries deoxygenated blood from the head.

renal vein (*n*) a vertebrate blood vessel that carries deoxygenated blood from the kidney; it joins the inferior vena cava (p.136).

iliac vein (*n*) a blood vessel in air-breathing vertebrates that carries deoxygenated blood from the hind limbs towards the heart. In mammals the **femoral vein** of the leg becomes the iliac vein; the two iliacs join to form the posterior vena cava (p.136).

red blood cell, erythrocyte (*n*) a disc-shaped cell found in vertebrate blood. It contains haemoglobin (p.133), a respiratory pigment which is responsible for the red colour. Red blood cells transport oxygen from the lungs to the tissues. They are elastic and easily distorted (↓) enabling them to pass through narrow capillaries. In most mammals they are biconcave, non-nucleated discs; in other vertebrates they are larger, oval in shape and have nuclei. There are about five million red blood cells in each cubic millimetre of human blood. Red blood cells are formed in red bone marrow; they have a relatively short life, about four months in man.

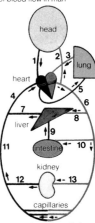

schematic representation of blood flow in man

trunk and lower limbs

1 superior vena cava
2 carotid artery
3 pulmonary vein
4 inferior vena cava
5 pulmonary artery
6 aorta
7 hepatic vein
8 hepatic artery
9 hepatic portal vein
10 mesenteric artery
11 inferior vena cava
12 renal vein
13 renal artery

━━ deoxygenated blood
━━ oxygenated blood
➡ flow to heart
--▶ flow from heart

red blood cell

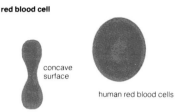

concave
surface

human red blood cells

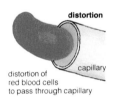

distortion

distortion of
red blood cells
to pass through capillary

capillary

distortion (*n*) the process of alteration of shape of an object. The structure is not destroyed and can regain its original shape; e.g. red blood cells (↑) are distorted as they pass through narrow capillaries.

haemolysis (*n*) the rupture of the cell membrane of a red blood cell (↑). The cell is destroyed and haemoglobin is released. This is caused, for example, by toxic chemicals, incompatible blood transfusions, or mechanical disruption.

plasma, blood plasma (*n*) the clear, fluid part of vertebrate blood. It is mainly water containing the soluble components of blood, e.g. plasma proteins (↓), inorganic ions, hormones; blood cells are suspended in it. It is formed from blood by centrifugation to remove all cells. Plasma can be clotted. *Contrast serum* (↓).

serum (*n*) the yellowish, watery fluid that separates from either blood or blood plasma (↑) when it clots. Its composition is the same as blood plasma except that it lacks the clotting constituents.

antiserum (*n*) type of serum (↑) produced from the blood of an animal that has been inoculated with a specific antigen (p.141) or pathogen (p.215), e.g. those causing tetanus and diphtheria. The blood and hence the antiserum contains antibodies to the antigen and gives immediate protection against these antigens, i.e. passive immunity.

blood sugar (*n*) glucose dissolved in blood plasma (↑). It is the final product of digestion of carbohydrate in mammals. Its concentration is kept fairly constant by the liver, which, under the influence of insulin, stores excess glucose as glycogen but which can, under the influence of glucagon and adrenalin, release it back to the blood if the glucose level in blood falls.

plasma proteins, blood proteins (*n*) proteins dissolved in vertebrate blood, e.g. antibodies and blood-clotting (p.140) substances, albumins.

blood clotting, blood coagulation (*n*) the conversion of whole, liquid blood or plasma (p.139) into a gel when blood vessels are injured. A blood clot (↓) is formed to prevent the escape of blood. In vertebrates the process involves the activation of soluble proteins, the clotting factors, in blood plasma. The first step is the liberation of thrombokinase (↓) from injured tissues or blood platelets. Thrombokinase, in the presence of Ca^{2+} ions, catalyses the formation of the enzyme thrombin from its precursor prothrombin. Thrombin catalyses the conversion of a soluble blood protein, fibrinogen, into a mass of insoluble fibrin threads which trap blood cells in its network. These threads shrink on drying forming a clot. Blood can be prevented from clotting by adding sodium citrate; this removes calcium ions from solution.

anticoagulant (*n*) a substance that prevents blood clotting (↑), e.g. **heparin**, found in all mammalian tissues, prevents formation of thrombin from prothrombin (↓).

plasmin (*n*) an enzyme found in normal blood; it dissolves a blood clot (↓) or thrombus (↓).

clot (*n*) a soft mass of material produced in a liquid by coagulation (↓), e.g. a blood clot. **clot** (*v*).

thrombus (*n*) a clot (↑) formed in a blood vessel or the heart. It may block the blood vessel thus causing a stroke or heart attack.

coagulation (*n*) the conversion of a liquid into a jelly-like solid state by chemical reaction, e.g. coagulation of blood during clotting (↑). **coagulate** (*v*).

prothrombin (*n*) a soluble blood protein. It is the inactive precursor of thrombin (↓).

thrombin (*n*) an enzyme formed from inactive prothrombin (↑) by the enzyme thrombokinase (↓). During blood clotting (↑) it converts the soluble blood protein fibrinogen into insoluble fibrin (↓).

thrombokinase (*n*) an enzyme liberated from both blood platelets and injured tissues. During blood clotting (↑) it converts prothrombin (↑) to thrombin in the presence of calcium ions.

thrombocyte (*n*) blood elements associated with blood clotting (↑). In most non-mammalian vertebrates they are spindle-shaped nucleated cells. In mammals they are the blood platelets. They disintegrate releasing thrombokinase (↑) to start blood clotting (↑).

fibrinogen (*n*) a blood protein (p.139), soluble precursor of fibrin (↓) into which it is converted by thrombin (↑).

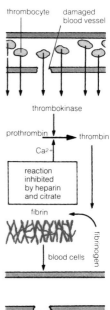

blood clotting

Blood groups	A	B	A, B	O
Antigens	A	B	A, B	—
Antibodies	anti-B	anti-A	—	anti-A anti-B

blood groups

Blood group of donor	Blood group of recipient		
	O	A	B	A, B (universal recipient)
O (universal donor)	√	√	√	√
A	X	√	X	√
B	X	X	√	√
AB	X	X	X	√

√ compatible transfusion   X incompatible transfusion

fibrin (*n*) an insoluble fibrous protein formed from its soluble precursor fibrinogen (↑) by the enzyme thrombin (↑). A clot (↑) is formed when blood cells are trapped in the fibrin network.

blood groups (*n.pl.*) classification of blood types in man into groups which can be mixed without agglutination (↓) of red blood cells. The ABO system is most commonly used. There are four main groups, A, B, AB, O, classified according to the antigen (↓) present on the surface of the red blood cells and the natural antibodies present in the plasma, e.g. a person of blood group A has antigen A on his red cells. He also has in his plasma antibodies to the red cell antigens not present on his red blood cells, i.e. in this case antibodies to antigen B. Antigens and antibodies (p.142) present in the other groups are shown in the diagram. The blood group of a patient must be determined before a blood transfusion is given. If blood from a person of group B (contains antigen B) is transfused into a person of group A (contains antibodies to B) then clumping or sticking together of the new red blood cells will occur due to the antibody-antigen reaction. Persons of blood group A can receive blood from groups A or O and can donate blood to recipients of type A or AB.

Rhesus factor, Rh factor (*n*) an antigen (↓) found on rhesus monkey and human red blood cells of Rh-positive but not Rh-negative individuals. Rh-negative individuals normally do not have antibodies (p.142) against this antigen but they may be acquired as a result of blood transfusion or pregnancy involving a Rh-positive foetus. Damage may then be caused to a Rh-positive foetus of any subsequent pregnancies.

agglutination (*n*) clumping or sticking together of cells when surface antigens on them interact with antibodies; e.g. when blood of group A is transfused into a person of blood group B, the antigens on the red cells are agglutinated by the antibodies present in the plasma of the recipient. This is used to identify blood groups. Similar reactions are used to identify bacteria. Substances producing agglutination are called **agglutinius**.

antigen (*n*) any foreign substance, usually protein or carbohydrate, that stimulates the recipient to produce a specific antibody (p.142) against it; the antibody coats the antigen and this results in its destruction and elimination from the body. *See p.222.*

antibody (*n*) protein found in animal blood, usually in response to an antigen, *see p.222.* The antibody combines with the antigen and this eventually leads to the destruction of the latter. Antibody formation is usually a defence mechanism, especially of vertebrates, against invasion by bacteria, fungi, parasites, some viruses.

blood film (*n*) thin smear of blood on a microscope slide. After staining, it can be used to detect the presence of parasites or determine blood count (↓).

blood count (*n*) the number of red and white cells in a defined volume of blood.

blood pressure (*n*) the pressure of blood in the main mammalian arteries. It varies according to the stage of the heartbeat from 80–120 mmHg and also with other factors, e.g. exercise, climate. It is measured using a sphygmomanometer (p.219).

pulse (*n*) a wave of increased pressure passing outwards from the heart along the arteries every time the ventricles contract. It is much faster than the rate of blood flow. The pulse is fainter as the distance from the heart increases; it is not found in the capillaries. The increased pressure can be felt at **pressure points**, i.e. where an artery lying close to the skin can be pressed flat against a bone. Usually the one in the wrist is used to measure the **pulse rate**, i.e. the number of pulse beats per minute.

lymphatic system (*n*) a system of thin-walled tubes that conduct lymph (↓) in vertebrates from the tissues into the circulatory system. The lymph capillaries, found in most tissues, join up to form larger tubes which finally link up with the venous system near the heart. Lymph capillaries are more permeable than blood capillaries; bacteria which cannot pass through blood capillary walls pass into the lymphatic system and are destroyed in lymph nodes (↓). In lower vertebrates lymph is pumped by **lymph hearts**, enlarged parts of lymphatic vessels with contractile, pulsating walls. In birds and mammals lymph flow is maintained partly by contraction of the lymph vessels and the squeezing of them by skeletal muscles and partly by the hydrostatic head of pressure developed by the heart. Valves in larger vessels maintain a uni-directional flow.

lymph (*n*) colourless fluid found in the lymphatic system (↑) of vertebrates. It is obtained from blood by filtration through capillary walls; red blood cells are absent and it contains less white cells and blood proteins but more lymphocytes than blood. *Compare tissue fluid* (p.11).

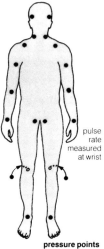

pulse
rate
measured
at wrist

pressure points

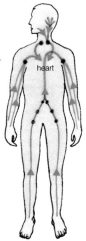

heart

lymphatic system
main drainage directions,
not actual lymph vessels

lymphatic vessels (*n*) thin-walled tubes in vertebrates that conduct lymph (↑). The smallest are the lymph capillaries. Many have lymph nodes (↓) and valves. Larger lymph vessels are called **lymph ducts**, e.g. the **thoracic ducts** in mammals and birds that drain lymph into the venous system at the base of the neck.

lacteal (*n*) a lymph capillary that drains the villi (p.106) of the intestine in vertebrates. Fat enters the lacteal, mixes with lymph and forms chyle (p.107).

lymph node, lymph gland (*n*) a structure found in larger mammalian lymph vessels (↑). It is less developed in birds and absent in other vertebrates. It is packed with lymphocytes which may become activated, if pathogens are present, to form plasma cells which produce antibodies (↑). As a result the nodes may become inflamed, a condition known as swollen glands.

white blood cell, leucocyte (*n*) a colourless cell found in blood and lymph of most vertebrates. It has a nucleus, shows amoeboid movement and lacks respiratory pigment. *Compare red blood cells* (p.138). It protects the body against pathogens and their toxins. There are two types: granulocytes (polymorphs p.144) that have granular cytoplasm; agranulocytes that lack granules in their cytoplasm, these are further divided into lymphocytes (↓) and monocytes (p.144). In man there are about 8000 white cells per mm^3 of blood.

agranulocyte (*n*) type of white blood cell (↑) in vertebrates that lacks granules in the cytoplasm, e.g. lymphocytes and monocytes. *Compare granulocyte* (↓).

granulocyte (*n*) type of white blood cell (↑) in vertebrates that has granules in the cytoplasm; all are polymorphs (p.144). *Contrast agranulocyte* (↑).

lymphocyte (*n*) agranular (↑) type of vertebrate white blood cell. They make up about 25% of the white cells in human blood. A lymphocyte has one large nucleus that almost fills the cell; there is little cytoplasm. It shows little amoeboid movement and does not act as a phagocyte. *Contrast polymorph* (p.144). In the presence of a specific antigen it can change into a plasma cell which has an extensive cytoplasm and produce antibodies, *see p.223*. Lymphocytes are produced in lymphoid tissues (↓).

lymphoid tissue (*n*) a vertebrate tissue that contains lymphocytes (↑). It is found in several parts of the body such as spleen (p.144), thymus and lymph nodes (↑).

lymphocyte

agranular leucocytes

monocyte

granular leucocyte

granular cytoplasm

polymorph

white blood cell

spleen (*n*) large, vascular organ in the abdominal mesentery of most vertebrates. It is connected to the circulatory system and is packed with lymphocytes (p.143) but is not connected to the lymphatic system. Functions include: destruction of old red blood cells; storage of red blood cells; production of antibodies by a mechanism similar to that of lymph nodes (p.143).

thymus, thymus gland (*n*) a vertebrate organ, usually found in the lower neck region, that is packed with lymphocytes (p.143). It functions as an endocrine gland with the role of switching on the immune system at birth. In most mammals it atrophies after puberty.

monocyte (*n*) a large, spherical, agranular white blood cell found in vertebrates. It is produced in lymphoid tissue (p.143), has an oval nucleus, shows amoeboid movement and acts as a phagocyte. Monocytes make up about 5% of human white blood cells. They are similar to macrophages (↓).

polymorph, polymorphonuclear leucocyte (*n*) an irregularly shaped, granular, white blood cell (p.143) found in vertebrates. There are several types; all are produced in bone marrow and act as phagocytes (↓). In some vertebrates, including man, the nucleus consists of several lobes. Polymorphs make up about 70% of human leucocytes. *See p.222.*

phagocyte (*n*) a cell that can ingest foreign particles by engulfing them into its own cytoplasm, a process known as phagocytosis (↓). Though many protozoans are phagocytes the term is particularly applied to the polymorphs, monocytes and macrophages found in blood.

phagocytosis (*n*) the engulfing or ingestion of foreign bodies by phagocytes (↑). The process involves the extension of pseudopodia around the foreign particle until it is finally enclosed in the cytoplasm of the phagocyte where it is digested.

polymorphonucleate (*adj*) describes a cell that has a nucleus with several distinct lobes, e.g. polymorph (↑).

macrophage (*n*) a large, motile phagocyte (↑) found widely in vertebrate tissue such as connective tissue, spleen, lymph nodes, bone marrow, liver. It takes up foreign particles from blood, lymph and damaged tissues *(see p.223).* Macrophages are practically identical to monocytes (↑).

reticulo-endothelial system (*n*) the system of

phagocytosis

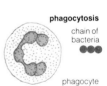

chain of bacteria

phagocyte

phagocyte engulfing bacteria

bacteria being digested

macrophages (↑) in vertebrates that are in contact with the blood of bone marrow, spleen and liver, or with lymph of lymph nodes. Macrophages remove foreign particles from the blood and lymph.

leukaemia (n) a disease in which there is an increase in the number of white blood cells. Excess of different types of white cell causes different types of leukaemia. As the number of white cells increases there is often a decrease in platelet and red blood cell production, resulting in anaemia (↓). Leukaemia may be fatal.

anaemia (n) a disease in which there is a reduction in the amount of haemoglobin in the blood; caused by faulty production of red blood cells and/or haemoglobin or by their destruction. Blood cannot transport enough oxygen to the tissues, resulting in weakness, pallor of the skin and reduced resistance to disease.

pernicious anaemia, megaloblastic anaemia (n) type of anaemia (↑) in which abnormal red cells (megaloblasts) are present in the blood. These cells can only transport very small amounts of oxygen.

haemorrhage, bleeding (n) loss of blood from injured blood vessels.

haemophilia (n) a hereditary disease of man. The genes responsible occur in both males and females but the symptoms of the disease are rarely shown by females. Blood cannot clot so **excessive** bleeding (↑) occurs.

arteriosclerosis (n) a condition of man in which artery walls become hardened and lose elasticity; the internal diameter is narrowed. If coronary vessels (p.136) are affected insufficient oxygen will reach heart tissues; this can lead to heart failure. Although controversial, there is some evidence to suggest that diets high in cholesterol (p.124) may contribute to this condition.

sclerosis (n) hardening of a tissue, e.g. by thickening of arteries; by lignification of plant cell walls.

varicose veins (n) a condition in man in which the veins, most often those of the legs, are abnormally dilated with blood. Valves in their walls do not function correctly.

phlebitis (n) a condition in man in which the wall of a vein becomes inflamed; a thrombus (p.140) forms inside.

thrombosis (n) the formation of a thrombus (p.140) in the blood vessels or heart of man.

septicaemia (n) a feverish condition in which pathogenic bacteria have entered the blood of animals from septic wounds or infected tissue.

breathing (*n*) process of pumping (1) air into and out of the lungs of terrestrial animals or (2) water over the gills of fish. Breathing involves inspiration (↓) and expiration (↓); it supplies oxygen and removes carbon dioxide. *Compare respiration* (p.148).

inspiration (*n*) part of the process of breathing (↑) that involves taking in from the surroundings either (1) air into the lungs of terrestrial animals or (2) water which is pumped through the gills of fish. **inspire** (*v*). *Compare expiration* (↓).

inhale (*v*) (1) to take in gases into the lungs of terrestrial animals; usually air during inspiration (↑) but it may involve other gases, e.g. anaesthetics like ether or chloroform; (2) to take in water for respiration in fish. **inhalation** (*n*).

expiration (*n*) part of the process of breathing (↑) that involves (1) removal of air from the lungs of terrestrial animals or (2) water which has passed through the gills of fish. **expire** (*v*). *Compare inspiration* (↑).

exhale (*v*) to expel gases from the lungs of terrestrial animals; usually this is air during expiration (↑) but it may involve other, obnoxious gases; (2) to expel water, used for respiratory purposes, that has passed through the gills of fish. **exhalation** (*n*).

respiratory movements (*n.pl.*) movements of part of an animal's body that result in air or water being passed to or from the respiratory organs (p.148). This supplies oxygen and removes carbon dioxide. The movements depend on the nature of the respiratory organs, e.g. in fish, muscles in the pharynx pump water over the gills; in terrestrial vertebrates contraction and relaxation of chest muscles draws air into or expels it from the lungs. In man this is aided further by the diaphragm.

respiratory centre (*n*) the part of the brain in vertebrates that controls respiratory movements (↑).

respiratory quotient, RQ (*n*) the ratio of the volume of carbon dioxide evolved relative to the volume of oxygen consumed in a given period of respiration. It depends on the substrate used for metabolism. For carbohydrates it is 1, for fats 0.7, for proteins 0.8.

breath (*n*) one single inspiration (↑) and expiration (↑).

respiratory rate (*n*) the number of breaths (↑) taken in one minute. In man this is about 18 to 20 but it can be increased, e.g. by a high concentration of carbon dioxide in the blood produced by strenuous exercise.

tidal air (*n*) the air entering and leaving the lungs during normal breathing (about 500 cm³ in man).

complemental air (*n*) the amount of air that can be forcibly inspired after normal inspiration to fill the lungs (about 2500 cm³ in man).

supplemental air (*n*) the amount of air that can be forcibly expelled after normal expiration (about 1000 cm³ in man).

residual air (*n*) air which remains in the lung and cannot be removed even by forced expiration (about 1200 cm³ in man).

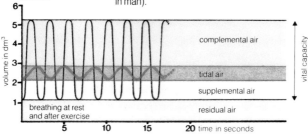

lung volumes during respiration at rest
lung volumes during respiration after exercise

breathing at rest and after exercise

complemental air

tidal air

supplemental air

residual air

vital capacity

volume in dm³

time in seconds

vital capacity (*n*) the amount of air moved by forced inspiration and forced expiration. It is the largest possible volume change of the lungs (about 4000 cm³ in man).

phlegm (*n*) mucus in which foreign matter, e.g. dust, has become trapped. It is found in the respiratory tract (p.148) of air-breathing vertebrates and is wafted by cilia up it to the pharynx and is coughed out via the mouth.

catarrh (*n*) fluid discharged from an inflammation of a mucous membrane, particularly of the human nose.

suffocation (*n*) prevention of breathing, especially in air-breathing vertebrates.

pneumonia (*n*) a disease, particularly of man, caused by bacteria (especially pneumococci); involves inflammation of the lungs.

pleurisy (*n*) inflammation of the membranes (pleura p.81) surrounding the lungs, especially in man. It often results in liquid being found in the pleural cavities.

bronchitis (*n*) inflammation of the bronchi (p.149), especially in man.

laryngitis (*n*) inflammation of the larynx (p.149), especially in man.

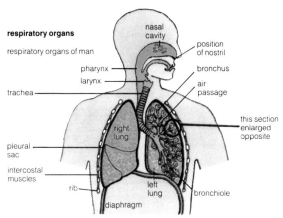

respiratory organs

respiratory organs of man

nasal cavity

position of nostril

pharynx

larynx

trachea

bronchus

air passage

right lung

this section enlarged opposite

pleural sac

intercostal muscles

rib

left lung

bronchiole

diaphragm

respiration (*n*) (1) the breathing process in which
oxygen is taken from the environment and carbon
dioxide is returned to it. In protozoans and lower
invertebrates oxygen diffuses in through the whole of
the organism and carbon dioxide diffuses out; in fish,
oxygen/carbon dioxide exchange occurs as water
passes through gills (p.34); in terrestrial vertebrates,
exchange occurs as air is pumped into and out of the
lungs (↓). This is often referred to as **external
respiration**. (2) the oxidative, enzyme-catalysed
process by which an organism obtains its energy from
foodstuffs. This is often referred to as tissue respiration
(p.150). **respire** (*v*).

respiratory pigment, blood pigment (*n*) coloured
substance which combines reversibly with oxygen and
so increases the amount of oxygen that can be carried
by the blood, e.g. haemoglobin, haemocyanin.

respiratory organs (*n*) specialised organs in animals,
e.g. trachea (↓) of insects, gills of fish (p.34), lungs (↓) of
terrestrial animals, that provide surfaces for exchange
of **respiratory gases** (i.e. oxygen and carbon dioxide)
between the organism and the environment.

lung (*n*) a respiratory organ (↑) found in air-breathing
terrestrial animals.

respiratory tract (*n*) the structures in an air-breathing
animal that conduct air from the nose to the alveoli (↓)
during external respiration (↑).

nares (*n*) paired openings from the exterior into the nasal cavity (↓) of vertebrates; in some, especially those with a snout or nose, e.g. man, they are called **nostrils**.

nasal cavity (*n*) a cavity in a vertebrate head that contains the olfactory organs (p.169). In some, e.g. fish, it is blind-ended but in others, e.g. man, it leads into the pharynx and is a passage for respiratory air.

trachea (*n*) (1) a tube of the **tracheal system** of insects to carry air from several surface openings to body tissues. It branches into thin-walled **tracheoles** across which exchange of oxygen/carbon dioxide takes place. (2) the windpipe of vertebrates. It is a single tube which carries respired air between the pharynx and bronchi. Its walls are supported by C-shaped plates of cartilage.

larynx (*n*) the upper end of the trachea (↑) in terrestrial vertebrates. It opens into the pharynx through the glottis (p.101). Its walls have plates of cartilage and muscles which close the glottis during swallowing. The larynx of amphibians, reptiles and mammals has vocal cords (↓). In man it is the **Adam's apple**.

vocal cord (*n*) fold of membrane in amphibians, reptiles and mammals projecting from larynx (↑) walls into the lumen. It vibrates as air is expelled, causing sound.

bronchus (*n*) one of the two tubes into which the trachea (↑) divides in terrestrial vertebrates. Each bronchus splits into smaller bronchi and finally into bronchioles (↓). It conducts air between the trachea and the bronchioles. Walls of the bronchus are supported by C-shaped cartilage plates. They are lined with ciliated epithelium that secretes mucus to trap dust; the cilia propel this as phlegm (p.147) to the mouth.

bronchioles (*n.pl.*) small, tubular branches of bronchi (↑) which end in numerous alveoli (↓); they conduct air between these two structures. Unlike bronchi the walls lack mucus-secreting glands or cartilage; smooth muscle in them can alter the size of the lumen.

alveolus (*pl.alveoli*) (*n*) one of the numerous, small, thin-walled, air-filled, membranous sacs at the terminal end of a bronchiole (↑) in vertebrate lungs. It is covered with capillaries and provides a large, moist surface through which exchange of oxygen/carbon dioxide takes place.

intercostal muscles (*n*) muscles lying between the ribs (p.90) of terrestrial vertebrates. They play a role in breathing (p.146).

costal (*adj*) concerned with vertebrate ribs (p.90).

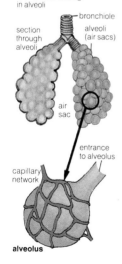

bronchiole
bronchioles ending in alveoli

bronchiole
section through alveoli
alveoli (air sacs)
air sac
entrance to alveolus
capillary network

alveolus

diaphragm (*n*) a sheet of tendon and muscle that separates the thoracic and abdominal cavities of mammals. When relaxed it is dome-shaped; when contracted it flattens, helping air movement into the lungs.

tissue, internal or **cell respiration** (*n*) the series of enzyme-catalysed chemical reactions in living organisms by which organic substances are oxidised to produce ATP (p.110). There are two types, see aerobic (↓) and anaerobic respiration (p.152).

aerobic respiration (*n*) type of tissue respiration (↑) in which organic substances are broken down in the presence of oxygen *(see diagram)*. More ATP is liberated in aerobic respiration than in anaerobic respiration (p.152). The first stage of both anaerobic and aerobic respiration of glucose is glycolysis (↓) which yields pyruvate. Under aerobic conditions the pyruvate produced enters the Kreb's cycle (↓) and in a series of reactions is broken down to carbon dioxide and reducing power in the form of NADH and NADPH (p.109). Oxidation of the latter by the respiratory chain (↓) liberates energy which is stored as ATP (p.110).

aerobe (*n*) organism that obtains energy by aerobic respiration (↑).

> **aerobic respiration**
> $$C_6H_{12}O_6 + 6O_2$$
> ↓
> $$6CO_2 + 6H_2O + \text{ENERGY}$$
>
> overall reaction of
> aerobic respiration

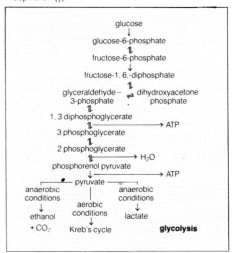

glucose
↓
glucose-6-phosphate
⇅
fructose-6-phosphate
↓
fructose-1, 6,-diphosphate
⇅
glyceraldehyde – ⇌ dihydroxyacetone
3-phosphate   phosphate
⇅
1, 3 diphosphoglycerate
⇅────────────→ ATP
3 phosphoglycerate
⇅
2 phosphoglycerate
⇅────────────→ H_2O
phosphorenol pyruvate
↓────────────→ ATP
●── pyruvate ──
anaerobic   anaerobic
conditions | conditions
              aerobic
 ↓   conditions   ↓
ethanol   ↓   lactate
+ CO_2 Kreb's cycle   **glycolysis**

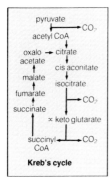

glycolysis, Embden–Meyerhoff pathway (*n*) an anaerobic sequence of enzymically-catalysed reactions by which glucose is converted to pyruvate (↓). Some energy is released and stored as ATP; the amount is much smaller than that produced via the Kreb's cycle (↓) and respiratory chain (↓). Glycolysis occurs in the cytoplasm of plants, bacteria and animals. It is the first stage in tissue respiration (↑), fermentation (p.152) and the release of energy in vertebrate muscle.

Kreb's, citric acid, tricarboxylic acid or **TCA cycle** (*n*) a cyclic series of enzymically-catalysed reactions by which pyruvate (↓), produced by glycolysis (↑), is oxidised to carbon dioxide and reducing power in the form of NADH and NADPH (p.109). Oxidation of the latter by the respiratory chain (↓) liberates energy which is stored as ATP. The cycle is also concerned with the final reactions of fat oxidation and the synthesis of some amino acids. It occurs in the mitochondria of eukaryotic cells (p.13).

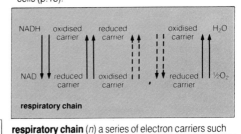

respiratory chain

respiratory chain (*n*) a series of electron carriers such as **cytochromes** which pass electrons from the NADPH or NADH produced in the Kreb's cycle (↑) to oxygen, forming water. Energy is released and used to synthesise ATP. The process occurs in the mitochondria of eukaryotic cells (p.13).

oxidative phosphorylation (*n*) synthesis of ATP during the passage of electrons down the respiratory chain (↑).

pyruvic acid (*n*) important compound in metabolism. It is produced mainly from glucose by glycolysis (↑). In aerobic respiration (↑) it is further metabolised by the Kreb's cycle (↑); in anaerobic respiration it is converted to either lactate or ethanol. Ionic form is **pyruvate**.

citric acid (*n*) important intermediate in the Kreb's cycle (↑); ionic form is **citrate**. It is found in high concentration in citrus fruit.

COOH
|
C=O   **pyruvic acid**
|
CH₃
formula of pyruvic acid

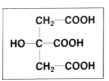

citric acid
formula of citric acid

anaerobic respiration (*n*) tissue respiration (p.150) in which energy as ATP is produced by breakdown of substances in reactions that do not use oxygen; e.g. breakdown of glucose to ethanol and carbon dioxide by yeasts and breakdown of pyruvate to lactate in vertebrate muscle. Less energy is produced than by aerobic respiration (p.150). All fermentations (↓) are types of anaerobic respiration. The facultative anaerobes (↓) and some tissues of multicellular organisms (e.g. vertebrate muscle) can respire anaerobically when there is insufficient oxygen for aerobic respiration.

anaerobe (*n*) organism that obtains energy by anaerobic respiration (↑).

fermentation (*n*) anaerobic respiratory (↑) process, especially of bacteria and yeasts, involving breakdown of organic substances, usually carbohydrate, to produce energy as ATP (p.110).

alcoholic fermentation (*n*) a fermentation (↑) found in certain yeasts and bacteria. Glucose is broken down by glycolysis (p.151) to ethanol and carbon dioxide.

lactic acid (*n*) an organic acid; ionic form is **lactate**. It is the end-product of anaerobic respiration (↑) of glucose in some bacteria (lactobacilli) and in vertebrate muscle under conditions of oxygen shortage.

oxygen debt (*n*) a state found in animals or parts of them when oxygen is temporarily lacking and therefore anaerobic respiration (↑) is being carried out; e.g. in vertebrate muscle pyruvate produced from glucose by glycolysis (p.151) is normally oxidised to carbon dioxide. During prolonged muscular exercise an oxygen debt is built up and the pyruvate reduced to lactate.

facultative anaerobe (*n*) organism that respires equally well in the presence or absence of molecular oxygen.

obligate anaerobe (*n*) organism that is unable to grow in the presence of molecular oxygen.

obligate aerobe (*n*) organism that requires molecular oxygen for respiration. Growth feeble or stops in absence.

oxidation (*n*) process by which (1) oxygen is added to a substance or more usually in biological systems hydrogen is removed; (2) electrons are removed from an ion, atom or group of atoms. *Compare reduction* (↓).

reduction (*n*) process by which (1) oxygen is removed from a substance or more usually in biological reactions hydrogen is added; (2) electrons are added to an ion, an atom or a group of atoms. *Compare oxidation* (↑).

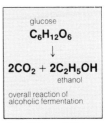

glucose

$$C_6H_{12}O_6$$

↓

$$2CO_2 + 2C_2H_5OH$$

ethanol

overall reaction of alcoholic fermentation

alcoholic fermentation

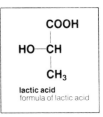

COOH
|
HO—CH
|
CH$_3$

lactic acid
formula of lactic acid

putrefaction (*n*) anaerobic decomposition of proteins, especially by bacteria.

homeostasis (*n*) the maintenance of a constant internal environment in animals; e.g. in mammals of constant osmotic pressure, pH, temperature and concentration of substances like glucose and dissolved gases.

osmoregulation (*n*) the maintenance of the correct levels of water, in animals so that the osmotic pressure of the internal environment (p.11) is kept constant. Excretory organs are usually associated with osmoregulation, e.g. the vertebrate kidney.

ionic regulation (*n*) control of level of individual ions.

excretion (*n*) the removal from an organism of the waste products of its metabolism. The main waste products are carbon dioxide, water and nitrogenous waste. In unicellular animals the products diffuse out through the cell membrane. In multicellular animals, carbon dioxide is usually excreted by respiratory organs (p.148) and nitrogenous waste (e.g. ammonia, urea and uric acid p.131) together with water is usually excreted by special excretory organs, e.g. nephridia (↓) of invertebrates, Malpighian tubules (↓) of insects and kidneys of vertebrates. **excrete** (*v*). *Compare defaecation* (p.108).

excreta (*n*) waste material formed by metabolism of an organism, e.g. carbon dioxide, ammonia, urea, uric acid, water. They are removed during excretion (↑). *Compare faeces* (p.108).

active transport (*n*) movement of dissolved substances across a cell membrane from a region of low concentration to one of high concentration. It uses energy (ATP) produced by metabolism.

passive transport (*n*) movement of dissolved substances across a cell membrane by diffusion (↓). It occurs from a region of high concentration to one of low concentration. It does not require energy.

diffusion (*n*) movement of a substance (usually a gas or dissolved solute; *see p.241*) from a region of high concentration to a region of low concentration.

filtration (*n*) process that separates solids from a liquid by passage of the liquid through a porous material.

ultra-filtration (*n*) filtration, under pressure.

nephridium (*n*) an organ of excretion (↑) found in many invertebrates. It is a tube into which waste products are passed for removal to the exterior.

Malpighian tubules (*n*) the excretory organs of insects.

kidney (*n*) one of two organs in vertebrates concerned with osmoregulation (p.153) and elimination of nitrogenous waste as urine. The structure varies in different species; it consists of many nephrons (↓), which open into the pelvis (↓), and their blood supply. In mammals it has an outer, dark red cortex (the **renal cortex**) in which the Malpighian bodies (↓) are concentrated. This surrounds an inner, paler **renal medulla**.

renal (*adj*) concerned with the kidney (↑), e.g. the renal artery supplies it with blood.

pyramid (*n*) a projection of the medulla into the pelvis (↓) of a vertebrate kidney (↑). In it collecting tubules join up to form ducts which are open at the apex of the pyramid to carry urine into the pelvis.

pelvis of kidney, renal pelvis (*n*) a cavity formed by the ureter (↓) where it joins the kidney. The pyramids (↑) open into it.

nephron (*n*) the structural and functional unit of a vertebrate kidney (↑). It consists of a Malpighian body (↓) and a uriniferous tubule (↓) leading from it.

Malpighian body, Malpighian corpuscle (*n*) part of the nephron (↑) of a vertebrate kidney. It consists of a Bowman's capsule (↓) and the glomerulus (↓) that it surrounds. High blood pressure in the glomerulus causes urea, glucose, soluble inorganic salts and water but **not** blood cells and proteins to be filtered through its capillary walls into the walls of the capsule. The filtrate passes into the uriniferous tubule (↓).

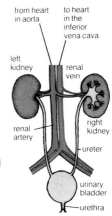

human urinary system

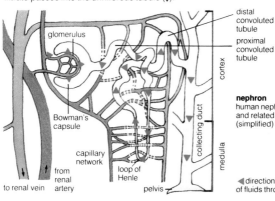

nephron
human nephron
and related structures
(simplified)

◀ direction of movement
of fluids through nephron

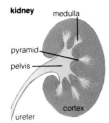

kidney

medulla

pyramid

pelvis

cortex

ureter

Bowman's capsule (*n*) part of the Malpighian body (↑) of a nephron (↑). It is the cup-shaped end of a uriniferous tubule (↓) that surrounds the glomerulus (↓).

glomerulus (*n*) the knot of capillaries surrounded by the Bowman's capsule (↑) in a vertebrate kidney.

uriniferous tubule, kidney tubule (*n*) the narrow, convoluted, tubular part of a nephron (↑) in a vertebrate kidney leading from the Bowman's capsule (↑) to the **collecting ducts** that carry urine away from the kidney. As the filtrate from the Bowman's capsule passes down it all the glucose, some of the inorganic ions (see osmo-regulation p.153) and in terrestrial animals most of the water are reabsorbed into the capillaries surrounding the tubule. Urea is not reabsorbed and together with the remaining water forms urine which drains into the Wolffian duct (↓) or into the ureter (↓). In mammals the uriniferous tubule is in three parts, the proximal con-voluted tubule, the **loop of Henle** and the distal convoluted tubule. The two convoluted parts are in the cortex, the loop of Henle lies mainly in the medulla.

ureter (*n*) a duct that carries urine from the kidney (↑) into the cloaca (↓) or into the urinary bladder (↓) of reptiles, birds and mammals. There are two, one from each kidney. In fish and amphibians this function is carried out by the Wolffian duct (↓).

Wolffian duct (*n*) a duct in fish and amphibians that carries urine from the kidney (↑); the testis also discharges sperm through it. *Compare ureter* (↑).

urinary bladder (*n*) a sac for temporary storage of urine in fish, amphibians and mammals. It is absent in birds and most reptiles. Urine is brought to it by a ureter (↑) or by a Wolffian duct (↑). In mammals its exit to the urethra (↓) is closed by a sphincter muscle.

urethra (*n*) a tube in mammals that carries urine from the urinary bladder (↑) to the exterior. In females it is shorter and leads to the vestibule. In males it is longer, leading through the penis; as it is joined by the vas deferens it discharges both urine and sperms to the exterior.

cloaca (*n*) a small chamber at the posterior end of most vertebrates. The kidneys, reproductive ducts and alimentary tract all empty their respective contents, i.e. urine, gametes and faeces, into it. The opening to the exterior is the **cloacal aperture**. The cloaca is not pre-sent in mammals, here the anal opening is separate from the genito-urinary (p.156) openings. *(See anus p.107.)*

genito-urinary system, urogenital system (*n*) the joint excretory and reproductive systems of vertebrates. Since the two systems are closely linked they are often considered together.

nephritis (*n*) inflammation of the kidney (p.154).

urine (*n*) a solution of waste metabolic products, produced in the kidney and excreted through the urethra or cloaca. It is mainly urea or uric acid (p.131) and sodium chloride with some other salts. By varying its amount and concentration, it serves to osmoregulate and ionically regulate the blood.

urination, micturation (*n*) emptying of the urinary bladder.

diuresis (*n*) increased production of urine (↑) by the kidney; occurs after drinking excess water or liquids.

diabetes (*n*) disease of man in which much urine (↑) is produced and excreted. **Diabetes mellitus** is caused by insufficient insulin production by the pancreas; glucose is present in the urine. **Diabetes insipidus** is caused by malfunction of the pituitary gland. Thirst is a symptom in both types of diabetes.

hair (*n*) (1) a cornified, thread-like outgrowth from the epidermis of mammals. Its length varies according to species and body region. Colour depends on amount of melanin (↓) present and also on the number of air bubbles; many air bubbles, e.g. in older humans, resulting in white hair. A hair consists of two parts: a shaft of dead cells, which projects above the skin surface; a root, embedded in the hair follicle (p.158) of the skin, where growth in length of hair occurs. Hairs protect against excess heat loss but they may also have a sensory function, e.g. whiskers (↓), or a protective function, e.g. spines. (2) in plants a thread-like outgrowth from an epidermal cell; it contains protoplasm, e.g. root hairs.

bristle (*n*) a short, stiff hair (↑) in mammals.

whisker (*n*) a long bristle (↑) or a stiff hair (↑) on the face with a sensory function, e.g. in mouse, cat or tiger.

fur (*n*) thick, soft, fine hair (↑) on the skin of some mammals, e.g. lion, tiger. It protects the body from excess heat loss.

mane (*n*) long hair (↑) found in the neck region of some mammals, e.g. horse, lion.

grooming (*n*) animal actions by which fur (↑) or feathers are cleaned. Animals either groom themselves or are groomed by others of the same species.

fur
fur covering body

whiskers
whisker

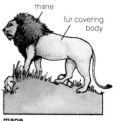

mane
fur covering body

mane

chromatophore (*n*) in animals, a cell that contains pigment. Of the three types, the most common, **melanophores,** contain **melanin** (a pigment). Different concentrations of melanin give a yellow or brown colour. Altering the pigment dispersion can result in rapid colour changes, e.g. in camouflage (↓) of chameleons and frogs.

camouflage (*n*) any means by which an animal merges with its surroundings in order to escape detection by predators.

cryptic colouration, protective colouration (*n*) the colour, and in multicoloured animals the arrangement of colours in such a way that the animal resembles its surroundings, e.g. stripes of tigers. *See camouflage* (↑).

albinism (*n*) a condition in mammals in which skin pigments fail to develop.

epidermis (*n*) in invertebrates an outer layer, usually one cell thick, that covers the body; it often secretes a cuticle. It is the outer layer of the skin (↓) of vertebrates. The epidermis in fish and amphibians is one layer thick; in terrestrial vertebrates it has several layers, the stratum corneum (↓), granular and the Malpighian layer (p.158). Feathers, scales, nails, claws, beaks and, in mammals, hair, sebaceous glands and sweat glands are all formed from it. It protects the body against injury and, in terrestrial animals, excess water loss.

skin (*n*) the external layer that covers the body and appendages of vertebrates. Connective tissue attaches it loosely to the body so it can move over structures lying under it. It is two layers thick consisting of an outer epidermis (↑) and an inner dermis (p.158). As well as protecting the body against injury and, in terrestrial animals, excess water loss it also contains receptors sensitive to touch, temperature, pain, etc., and so acts as a sense organ. Fish and amphibian skin has glands that secrete mucus for protection; that of fish and some reptiles has protective scales. In warm-blooded animals (i.e. mammals and birds) skin structures like fur, hair and feathers help to control body temperature.

stratum corneum, horny or **cornified layer** (*n*) outer layer of the epidermis (↑) in terrestrial vertebrates. It consists of dead keratinised (p.158) cells which are continually worn off and replaced by cells from the Malpighian layer (p.158). It is especially thick on the palms of hands and soles of feet. It protects the body against water loss, entry of pathogens and ultra-violet rays.

Malpighian layer (*n*) inner layer of the epidermis (p.157) of terrestrial vertebrates; below it lies the dermis (↓). Its living cells divide frequently but gradually lose their protoplasm, become impregnated with keratin (↓) and are pushed into the stratum corneum (p.157) to replace cells which have been rubbed off. Cells of the Malpighian layer often contain the pigment melanin which gives the skin its colour and affords protection against ultra-violet rays of the sun.

keratinisation, cornification (*n*) a process by which epidermal cells accumulate keratin (↓), leaving a dead, horny cell; this occurs in the outer cells of the epidermis (p.157) in vertebrates and in the formation of hair, wool, feathers, horny scales, hooves, nails, claws, beaks, horns.

keratin (*n*) a fibrous, sulphur-containing protein deposited in epidermal cells *(see keratinisation (↑))*.

dermis (*n*) the inner layer of the skin (p.157) in vertebrates. It lies below the epidermis (p.157) and is much thicker than it. It consists of connective tissue with collagen fibres providing strength and flexibility, fat cells for insulation and blood vessels for temperature regulation; lymph vessels, muscles, nerves, sebaceous glands (↓), sweat glands (↓) and hair follicles (↓) are also found in it. In some species it contains scales or bone.

hair follicle (*n*) a sheath formed by the epidermis projecting into the dermis in mammalian skin (p.157). It surrounds the root and shaft of a hair (p.156). Attached to it is an erector muscle (↓). A duct from a sebaceous gland (↓) opens into it.

erector muscle (*n*) a muscle attached to a hair follicle (↑). It contracts causing the hair to stand upright.

sebaceous gland (*n*) a gland that secretes sebum (↓) into the hair follicle (↑) in mammals. It is formed from the epidermis (p.157) but projects into the dermis (↑).

sebum (*n*) a fatty secretion of the sebaceous glands (↑) which waterproofs both hair and skin.

sweat glands (*n*) a coiled gland in mammalian skin (p.157) that secretes sweat (↓). Sweat glands are found throughout most of the skin in man and higher primates, other mammals may have fewer. They are formed from the epidermis (p.157) but project into the dermis (↑). A duct carries the sweat formed to the skin surface where it opens at a sweat pore(↓).

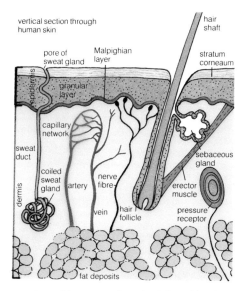

vertical section through human skin

hair shaft

pore of sweat gland

Malpighian layer

stratum corneaum

epidermis

granular layer

capillary network

sweat duct

coiled sweat gland

dermis

artery

nerve fibre

vein

hair follicle

sebaceous gland

erector muscle

pressure receptor

fat deposits

sweat pore (*n*) a very small opening in the epidermis (p.157) where the duct from a sweat gland (↑) opens on to the skin surface. Sweat exudes (↓) from it.

sweat (*n*) a product of mammalian sweat glands (↑) consisting mainly of a dilute solution of sodium chloride with a little urea. Evaporation of it from the skin surface helps cool the body; it plays a small part in osmoregulation (p.153). **to sweat** (*v*).

exude (*v*) to come out of a surface slowly, e.g. sweat (↑) as it emerges in small drops from a sweat pore (↑) on to the skin surface.

perspire (*v*) to exude (↑) small quantities of sweat (↑).

cutaneous (*adj*) concerned with the skin (p.157) of animals.

subcutaneous, sub-dermal (*adj*) concerned with structures lying under the skin (p.157) or dermis (↑) of animals.

squamous (*adj*) (1) consisting of scales, e.g. the skin of a lizard. (2) describes thin, flat, scale-like cells, e.g. squamous epithelium.

dermatitis (*n*) inflammation of the surface of the skin (p.157).

co-ordination (*n*) control of several structures or processes so all work together for the good of the whole;
e.g. animals co-ordinate body functions by the
combined action of the nerves of the nervous system (↓)
and hormones of the endocrine system (p.180).

nervous system (*n*) a network of cells present in all
multicellular animals. It is absent in unicellular
organisms (e.g. protozoa) and plants. Changes in the
internal and external environment are detected by
sense organs, tissues or cells. These generate nerve
impulses which are transmitted by afferent nerves, in
most animals to a CNS (↓). The CNS initiates impulses
which travel in efferent nerves to appropriate effectors
(e.g. muscle), to produce and co-ordinate suitable
responses. Nerves are made up of nerve cells called
neurons, linked by synapses (p.165). Afferent neurons
start at sense organs, efferent neurons terminate at
effectors, e.g. muscles. The simplest type of nervous
system is the nerve net (↓); more advanced systems
have nerve tracts and ganglia and have developed into
two interrelated parts, the central nervous system (↓)
and the peripheral nervous system (↓).

nerve net (*n*) the simplest type of nervous system (↑)
found, e.g. in coelenterates (p.22). A centre for co-
ordination (↑) of stimuli is absent. It consists solely of a
network throughout the tissues of simple nerve cells
lacking the separate axons and dendrons of the
neurons (p.164) of higher animals.

nerve cord (*n*) a strand of nervous tissue of the CNS (↓)
of an invertebrate running ventrally along the body and
bearing a ganglion in each segment. The analogous
structure in vertebrates is the spinal cord (↓).

ganglion (*n*) (*pl.ganglia*) a swollen region of a nerve
cord (↑) of invertebrates or nerve of vertebrates. It
contains many cell bodies of neurons and synapses.

plexus (*n*) a complex of nerves and ganglia (↑). Each
plexus is normally connected to several organs that
must have co-ordinated activities, e.g. in man the **solar
plexus** supplies nerves to the abdominal organs.

central nervous system, CNS (*n*) in animals with more
advanced nervous systems (↑), the part which co-
ordinates activities. In most invertebrates, e.g. Annelida
and Arthropoda, it consists of a ventral nerve cord (↑)
containing a swollen region, the ganglion, in each
segment. Head ganglia are largest and predominate;

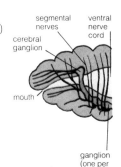

nerve cord
anterior part of
nervous system
of earthworm –
viewed from the side

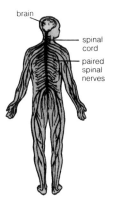

peripheral nervous system

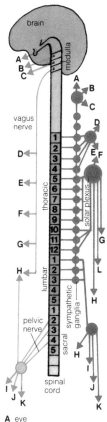

A eye
B lacrymal gland
C salivary gland
D heart
E lungs
F stomach
G small intestine
H large intestine
I kidney
J bladder
K sex organs
L adrenal gland

     parasympathetic
     nervous system
     sympathetic
     nervous system

they are equivalent in position to the brain of vertebrates. The CNS of vertebrates is a dorsal, hollow, thick-walled tube, i.e. the brain and spinal cord lying within the skull and backbone. Most of the synapses between vertebrate neurons occur in it and so it co-ordinates activities; impulses from sense organs normally travel to it and impulses to muscles and glands from it. *Compare nerve net* (↑).

peripheral nervous system (*n*) in animals with advanced nervous systems (↑) the parts not included in the CNS (↑). It consists of **peripheral nerves**, most of which lack synapses. They run from receptors to the CNS and from the CNS to glands and muscles.

autonomic nervous system (*n*) part of the nervous system (↑) that is concerned with visceral activities that are usually under involuntary (autonomic) control. It is not dependent on external stimuli. Simple autonomic systems are found in some invertebrates; vertebrate systems are more developed and divided into the sympathetic (↓) and parasympathetic (↓) nervous systems.

sympathetic nervous system (*n*) part of the autonomic nervous system (↑) of vertebrates such as birds and mammals. It is concerned mainly with homeostasis (p.153), e.g. it regulates heat loss and also prepares the animal for emergencies by causing an increase in heartbeat and release of adrenalin. Nerves from this system, unlike those from the parasympathetic system (↓), supply the skin and limbs.

parasympathetic nervous system (*n*) part of the autonomic nervous system (↑) of vertebrates like birds and mammals. It promotes functions like digestion and glandular secretion. In organs connected to both a sympathetic (↑) and parasympathetic system (as most are) stimulation of the parasympathetic system tends to restore normal function, i.e. it antagonises the emergency functions brought about by sympathetic stimulation, e.g. it decreases heartbeat.

spinal cord (*n*) part of vertebrate CNS (↑) within the vertebral column, surrounded by membranes. It contains white matter outside and an H-shaped core of grey matter surrounding a small, central canal. Paired spinal nerves (↓) start from it at nerve roots (p.162). Reflexes in it carry out simple co-ordination, e.g. of limbs.

spinal nerves (*n*) a series of paired, peripheral nerves arising from the spinal cord (↑) of a vertebrate at nerve roots (p.162). There are 31 pairs in man.

cranial nerves (*n*) a series of paired, peripheral nerves that emerge from the brain of vertebrates and supply sensory and/or motor nerves to the head, neck and viscera. In mammals there are 12 pairs. Included are: **olfactory nerves** and **optic nerves** supplying the nose and eye respectively; **facial nerves** supplying the face, salivary glands and taste buds of the tongue; **auditory nerves** to the ears; **vagus nerves** connected to muscles of the pharynx, larynx, oesophagus, stomach, heart and glands in the thoracic and visceral cavities.

nerve roots (*n*) found only in vertebrates, they are the points at which cranial (↑) and spinal nerves (p.161) start in the brain and spinal cord respectively. Each nerve has a **dorsal root** and a **ventral root**.

receptor (*n*) the part of an animal that responds to stimuli from inside or outside the body by sending impulses through nerve fibres to which it is connected. Receptors can be dispersed, e.g. those in the skin responding to touch, pain, etc. or they can be concentrated in a sense organ, e.g. the rods and cones of the eye.

sensory, afferent nerve (*n*) a peripheral nerve consisting of the nerve fibres of sensory neurons (↓).

sensory, afferent neuron (*n*) a nerve cell with nerve fibres that connect with and conduct impulses from a receptor (↑) to the CNS.

motor, efferent nerve (*n*) a peripheral nerve consisting of nerve fibres of motor neurons (↓).

motor, efferent neuron (*n*) a nerve cell with a fibre that connects with and conducts impulses from the CNS to an effector, e.g. muscle. Those conducting impulses to glands are often called **secretory neurons**.

mixed nerve (*n*) a peripheral nerve containing both sensory (↑) and motor (↑) nerve fibres.

effector (*n*) a cell or organ in multicellular animals that is specialised to carry out an action in response to a stimulus from the CNS, e.g. muscles, glands.

association, internuncial neuron (*n*) a neuron that lies between a sensory (↑) and a motor (↑) neuron. Its dendrites are linked by synapses to a sensory neuron axon; its axon is linked by synapses to dendrites of motor neurons.

stimulus (*n*) any change in the internal or external environment of an organism which is large enough to provoke a receptor to initiate a nerve impulse, e.g. touching a hot object gives a sensation of pain.

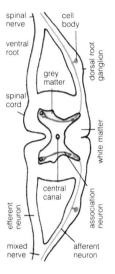

spinal nerve
connections of paired spinal nerves to the spinal cord

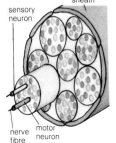

spinal nerve
cross section of spinal nerve

reflex arc
simple reflex arc

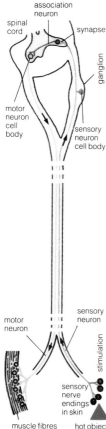

association neuron

spinal cord

synapse

ganglion

motor neuron cell body

sensory neuron cell body

motor neuron

sensory neuron

stimulation

sensory nerve endings in skin

muscle fibres contract and withdraw hand from hot object

hot object

threshold, minimal stimulus (*n*) the smallest stimulus (↑) that causes a receptor to initiate a nerve impulse.

impulse (*n*) the 'message' conducted along a nerve fibre. It consists of a wave of electrical changes associated with its membrane. It travels rapidly and is faster in large diameter nerves and those with a medullary sheath. After one impulse has passed a particular point on the nerve fibre, it cannot conduct again for a short time, called the **refractory period**.

latent period (*n*) time between the application of a stimulus and the beginning of the effector response; used in relation to tissues. *Compare reaction time* (↓).

reaction time (*n*) time between the application of a stimulus and the beginning of the effector response. It tends to be used in relation to responses of whole animals or plants. *Compare latent period* (↑).

response (*n*) change in the activity of part or a whole organism as a result of a stimulus.

inhibition of the nervous system (*n*) prevention of the normal actions of an effector in response to impulses received from the CNS, e.g. when a muscle is stimulated to contract, inhibition of the motor neurons of the appropriate antagonistic muscles causes their simultaneous relaxation.

interpretation (*n*) the way an animal (a) identifies a stimulus (b) decides, as a result of previous knowledge, the possible effect of it (c) prepares to act in response.

voluntary actions (*n*) actions controlled by an animal's brain. *Compare involuntary actions* (↓).

involuntary actions (*n*) actions not controlled by an animal's brain, e.g. reflex action (↓). *Compare voluntary action* (↑).

reflex action (*n*) behaviour in most animals where one simple stimulus causes a particular, simple response, e.g. the knee jerk. The response is immediate because there is a reflex arc (↓) to conduct impulses. This is an involuntary action (↑).

reflex arc (*n*) an inborn nervous pathway. The simplest involves three neurons (sensory, association and motor neurons) – *see diagram*; e.g. a hand placed on a hot object is quickly withdrawn. Note that even simple re-flexes involve synapses in the CNS so that they can be co-ordinated with other events by nervous inhibition (↑).

nervous tissue (*n*) tissue of the nervous system, i.e. neurons (nerve cells) (p.164) plus associated tissues.

nerve tract (*n*) a bundle of nerve fibres (↓) in the CNS; all the fibres start and end at similar connections.

nerve (*n*) a bundle of parallel nerve fibres (axons + dendrons) (↓), their associated blood vessels and supportive tissue, enclosed by a sheath of connective tissue, e.g. cranial nerves, spinal nerves (p.161). Each fibre conducts nervous impulses independently. There are sensory, motor and mixed nerves (p.162).

innervation (*n*) the type and distribution of nerves (↑) in a tissue, organ or organism.

neural, neuro (*adj*) concerned with the nervous system.

neuron, neurone (*n*) the basic element of the nervous system (p.160) in all animals. It is a cell specialised to conduct nerve impulses. It is a cell body (↓) with a nucleus surrounded by cytoplasm and projecting from this body are various processes, normally a single long axon (↓) and many dendrons (↓). (Those with one axon and several dendrons are **multipolar neurons**; those with one axon and one dendron are **bipolar neurons**.) Impulses pass from neuron to neuron via synapses (↓).

nerve cell (*n*) (1) the whole neuron (↑). (2) occasionally used to described only the cell body (↓) of a neuron.

cell body, perikaryon, cyton (*n*) part of a neuron (↑). It is the cytoplasm surrounding the nucleus. The term excludes the dendrons (↓) and axon (↓). Cell bodies are found in the grey matter of the brain and spinal cord and in the ganglia.

dendron (*n*) one of several cytoplasmic processes arising from the cell body (↑) of a neuron. It carries impulses towards the cell body. Dendrites (↓) branch from it. Dendron length varies in different types of neuron.

dendrite (*n*) (1) a terminal branch of a dendron (↑). (2) a dendron plus its terminal branches. Dendrites receive impulses from axons (↓) of other neurons.

axon (*n*) the long, normally single, cytoplasmic process arising from the cell body (↑) of a neuron. It conducts impulses away from the cell body; these pass either to another neuron or to an effector, e.g. muscle. The axon is branched at its end and is often covered with a myelin sheath (↓).

giant axon, giant fibre (*n*) axon (↑) with very large diameter, found in many invertebrates and some vertebrates. Conduction of impulses is very quick for rapid co-ordination of body movement.

nerve fibre (*n*) the axon (↑) of a neuron. If it is covered

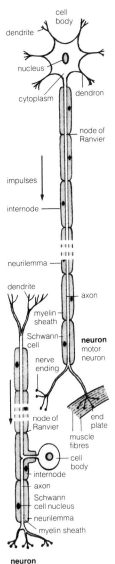

neuron

with a myelin sheath (↓) it is a **medullated nerve fibre**; if it is not it is a **non-medullated nerve fibre**.

medullary sheath, myelin sheath (*n*) a layer of myelin (↓) surrounding axons (↑) of most vertebrate and some invertebrate neurons. It is not continuous; regions covered with myelin are **internodes**, those not are **nodes** – the nodes of Ranvier (↓). A medullary sheath insulates the axon. It is bounded by the neurilemma (↓).

myelin (*n*) organic substance found in the myelin sheath (↑) of nerve fibres. It is composed of a fat and protein.

Schwann cell (*n*) cell of vertebrate nervous system that forms the myelin sheath (↑).

neurilemma (*n*) thin, outer cover of peripheral nerves.

node of Ranvier (*n*) a region of a medullated nerve fibre (↑) where the myelin sheath (↑) is interrupted. They occur at regular intervals along the fibre. Lymph provides nourishment at these points.

nerve ending (*n*) the peripheral ending of a nerve fibre. At this point in a sensory fibre impulses are initiated, or in a motor fibre they are terminated. Nerve endings may be simple, branched ends of the fibre or they may be organised into an end-plate (↓) of a muscle.

end-organ (*n*) structure at the end of a peripheral nerve. It may be a receptor or sensory neuron to receive stimuli from the environment or an end-plate (↓).

end-plate (*n*) the plate-like ending of a motor neuron (p.162) where it is attached to a muscle fibre.

synapse (*n*) the junction between the terminal branches of the axon (↑) of one neuron and either the dendrites of another neuron or an effector organ, e.g. muscle. One axon can form synapses with dendrites from several different neurons. In higher animals impulses travel through synapses only from axon to dendrites. In such animals, synapses between two neurons occur mainly in the CNS. Transmission of a nerve impulse from axon to dendrite or effector involves the release of a neuro-transmitter, in higher animals usually acetylcholine (↓).

acetylcholine (*n*) the most common of the chemical substances (i.e. **neurotransmitters**) involved in the transmission of nerve impulses across a synapse.

nerve regeneration (*n*) regrowth of nerves (↑) after they have been severed; only medullated nerves can do this.

multiple sclerosis (*n*) hardening of spinal cord or brain tissue. Nervous tissue (p.163) is destroyed causing paralysis and finally death.

$$CH_3-\overset{\overset{\textstyle O}{\|}}{C}-O-CH_2-CH_2-N\equiv(CH_3)_3$$

acetylcholine

formula of acetylcholine

brain (*n*) the co-ordination centre of the nervous system in higher animals. It is found at the anterior end; lower animals without an anatomically distinct head region do not have a brain. Some invertebrates, e.g. Arthropoda, Annelida, have enlarged ganglia (p.160) at the anterior end of their nerve cord which act as a primitive brain. The brain in vertebrates is divided into three regions, the fore-brain (↓), mid-brain (↓) and hind-brain (↓).

brain stem (*n*) the vertebrate brain (↑) excluding the cerebral hemispheres (↓) and cerebellum (↓), i.e. it is the mid-brain (↓) and the medulla oblongata (↓).

hind-brain (*n*) the cerebellum (↓) and medulla oblongata (↓) of a vertebrate brain. It lies between the spinal cord and the mid-brain (↓).

medulla oblongata, medulla (*n*) the posterior part of the hind-brain (↑); it is continuous posteriorly with the spinal cord. It has grey matter on the inside and white matter on the outside; *compare cerebellum* (↓). It regulates automatic, vital processes like respiration, heartbeat, dilation of blood vessels, and reflexes like sneezing and swallowing.

cerebellum (*n*) an outgrowth from the hind-brain (↑) of a vertebrate. Compared to other brain structures it is small in reptiles and amphibians, larger in mammals and very large in birds and fish. In mammals it has white matter on the inside and grey matter on the outside; *compare medulla oblongata* (↑). Its lobes are connected by the pons Varolli (↓). It co-ordinates complex muscular movements and maintains a sense of balance.

pons Varolli, pons (*n*) a tract of nerve fibres between the medulla oblongata (↑) and the mid-brain (↓). It conducts impulses between the individual lobes of the cerebellum (↑) and also between the cerebellum and the medulla oblongata and the cerebrum (↓). It also helps in the regulation of respiration.

mid-brain (*n*) region of a vertebrate brain between the fore-brain (↓) and hind-brain (↑). It extends from the lower surface of the cerebrum (↓) to the pons (↑). It is a short, narrow, thick-walled, tubular structure particularly concerned with sight and hearing.

optic lobes (*n*) paired lobes lying on either side of the brain stem of the mid-brain (↑) of vertebrates. Optic lobes are mainly concerned with vision.

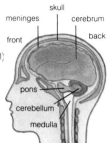

brain
location of the brain within the skull of man

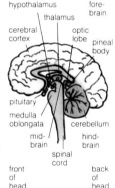

brain
longitudinal section through one hemisphere of human brain

fore-brain (*n*) the part of a vertebrate brain lying anterior to the mid-brain (↑). It is concerned with the senses of smell, vision (especially in mammals) and maintenance of equilibrium. In higher vertebrates it is greatly developed *(see cerebrum, thalamus, hypothalamus* (↓)*)* and is concerned with conscious actions.

optic chiasma (*n*) the point where the optic nerves cross over. In most vertebrates all the nerve fibres of the right optic nerve cross to the left side of the brain and vice versa. In mammals only half cross over.

thalamus (*n*) a large mass of grey matter in the fore-brain (↑) of higher mammals including man. It relays sensations (e.g. of heat, cold, pain) to the cerebrum (↓).

hypothalamus (*n*) a structure found in the fore-brain (↑) of higher mammals below the thalamus (↑) and above the pituitary gland (↓). It is the main co-ordinating centre for visceral functions; it regulates water balance and body temperature and controls sleep, appetite and aggression. It also controls the pituitary gland and so co-ordinates neural and hormonal functions.

pineal body, – gland (*n*) small outgrowth from vertebrate fore-brain (↑); it may have an endocrine function.

pituitary gland, – body (*n*) an endocrine gland found below the hypothalamus (↑) to which it is attached by a short stalk. It has two lobes that secrete several hormones; those of the anterior lobe include ACTH and growth hormones; those of the posterior lobe include ADH and oxytocin (p.181). It controls the activities of most other endocrine glands. The pituitary gland itself is controlled by the hypothalamus.

cerebrum (*n*) an outgrowth from the fore-brain (↑) of vertebrates. In lower vertebrates it is concerned with smell. In higher vertebrates it is separated into two halves by a fissure and has developed co-ordination functions; some, e.g. mammals, have hemispherical shaped halves called **cerebral hemispheres**. In mammals further development results in it being the largest brain structure. It controls most of the animal's activities, it interprets stimuli from receptors and causes voluntary actions in effectors. In man it is also the seat of mental activities, e.g. reasoning, emotion. The various functions of the body are associated with particular regions of the cerebrum. The outer region, the cerebral cortex (p.168), is grey matter; it surrounds white matter, the cerebral medulla.

cerebral cortex (*n*) the grey matter forming the outer layer of the cerebral hemispheres of the brains of reptiles, birds and mammals. It surrounds the white matter, the **cerebral medulla**. Many synapses occur in it and it is the main region for integration of the nervous system. It is extensive in mammals where it is much folded; the more folds the greater the intelligence.

cerebral (*adj*) concerned with the brain.

olfactory lobes (*n*) a pair of outgrowths from the terminal end of the fore-brain in lower vertebrates or from the cerebral hemispheres of higher vertebrates. They are concerned with smell (↓), being less developed in animals, e.g. man where a sense of smell is less important. The olfactory nerve (↓) runs to them from the olfactory organ (↓).

grey matter (*n*) a type of nervous tissue present in a vertebrate CNS. Its nerve fibres do not have a myelin sheath (p.165) and therefore are grey in colour. In the spinal cord, it is found internal to the white matter (↓); in the cerebellum and cerebral hemispheres it is external to it. It consists of cell bodies, dendrites and the terminal ends of axons plus their supportive tissues and blood vessels. Since most synapses are found here it forms regions of the CNS where co-ordination occurs.

white matter (*n*) a type of nervous tissue present in a vertebrate CNS. Its nerve fibres have a myelin sheath (p.165) and therefore are white. It lies outside the grey matter (↑) in the spinal cord, but internal to it in the cerebellum and cerebral hemispheres. It contains tracts of nerve fibres plus supportive tissues and blood vessels.

cerebrospinal fluid (*n*) liquid which is present in the cavities of a vertebrate CNS. It nourishes the nervous tissue and protects it from mechanical injury.

meninges (*n*) membranes covering the vertebrate brain and spinal cord. Except for fish, three are present: (1) the tough, outer **dura mater** (2) the delicate **arachnoid** (3) the delicate, inner **pia mater**.

meningitis (*n*) inflammation of the meninges (↑).

electroencephalogram, EEG (*n*) a record of the small electrical impulses produced by the brain. It is obtained using an instrument called an **electroencephalograph**.

sense organ (*n*) in animals, a collection of sensory receptors and associated tissues specialised to respond to a particular stimulus, e.g. light, taste. See olfactory organ (↓), taste buds (↓), ear (p.176), eye (p.170).

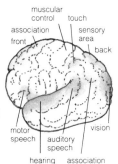

brain
localisation of functions in cerebral cortex in man

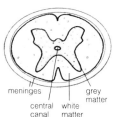

grey matter
white matter
spinal cord showing grey and white matter

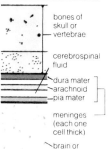

meninges
protection of brain and spinal cord

exteroceptor (*n*) in animals, a receptor (p.162) or a sense organ (↑), e.g. ear, eye, that detects stimuli coming from outside the body. *Contrast interoceptor* (↓), *proprioceptor* (↓).

interoceptor (*n*) a receptor in or on an internal organ in an animal that detects stimuli arising inside the body. *Compare exteroceptor* (↑).

proprioceptor (*n*) in animals a receptor that detects position and movement, e.g. (a) those found in tendons, muscles and joints that are stimulated by stretching or contraction of muscles, (b) those in the inner ear that are stimulated by movement.

sense (*n*) the ability of an animal to receive sensations, i.e. to be aware of its environment. Vertebrate senses include sight, hearing, smell, taste, touch.

sensory (*adj*) concerned with the reception of stimuli, e.g. a sensory nerve conducts impulses from sense organs to the CNS.

smell (*n*) in animals the sense by which gases and vapours are perceived. It is detected by olfactory organs (↓) which send impulses along the olfactory nerve (↓) to the brain for interpretation.

olfactory (*adj*) concerned with the sense of smell, e.g. the olfactory nerve (↓), olfactory organs (↓).

olfactory nerve (*n*) a sensory, cranial nerve (p.162) connected to mucus membrane of olfactory organ (↓).

olfactory organs (*n*) the organs of smell (↑). In invertebrates they are found in various positions, e.g. on the antennae of arthropods. In terrestrial vertebrates they are areas of sensory epithelium in the upper region of the nasal cavity

taste (*n*) in animals, the sense (↑) by which dissolved substances are detected by the taste buds (↓) of the tongue.

taste buds (*n*) receptors which are sensitive to taste (↑). They are groups of sensory cells found over the surface of the tongue in mammals and over the tongue and buccal cavity of other vertebrates. There are four types of taste bud, classified according to the type of stimulus to which they are sensitive: bitter, salt, sour, sweet. Areas of the tongue differ in sensitivity.

touch (*n*) in animals, the sense by which contact with material substances is perceived. It is detected by pressure and tactile receptors found below the epidermis.

tactile (*adj*) concerned with the sense of touch (↑).

attached end

bitter

sour

sweet

taste
human tongue showing areas most sensitive to different tastes

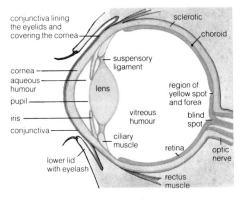

conjunctiva lining the eyelids and covering the cornea

sclerotic

choroid

suspensory ligament

cornea

aqueous humour

pupil

iris

conjunctiva

lens

region of yellow spot and forea

vitreous humour

blind spot

ciliary muscle

retina

optic nerve

lower lid with eyelash

rectus muscle

eye
longitudinal section
of mammalian eye

eye (*n*) the sense organ of sight (↓). It is an organ which responds to the stimulus of light. Very simple eyes are found in many invertebrates. More complex eyes are found in crustaceans, insects *(see compound eye,* p.40*)*, cephalopod molluscs and vertebrates.

photoreceptor (*n*) a receptor that is sensitive to light, e.g. special cells in the epidermis of invertebrates like earthworms; the eye (↑) of vertebrates.

sight (*n*) a sense found in arthropods and vertebrates. In vertebrates light stimulates photoreceptors (↑) to send impulses along the **optic nerve** to the CNS so that the animal is aware of its physical surroundings.

vision (*n*) the process by which an animal uses its eyes (↑) to become aware of its physical surroundings. Different animals have different forms of vision. For example, mammalian herbivores need a large field of vision so they can detect predators; they have **monocular vision** in which each eye is set at the side of the head so that an object is seen with only one eye. In contrast, **binocular vision** is found in primates and in some other vertebrates, especially predators, that need to judge distances. A pair of eyes is found at the front of the head, the image of the object falls on both retinas (↓). The slightly different pictures of the object obtained by the two eyes looking from different angles allow solid objects to be seen stereoscopically, i.e. in three dimensions.

visual, optic, ocular (*adj*) concerned with vision (↑) or the eye (↑).

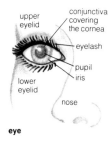

upper eyelid

conjunctiva covering the cornea

eyelash

pupil

iris

lower eyelid

nose

eye

right mammalian eye
– external features

ophthalmic (*adj*) concerned with the eye (↑).

orbit (*n*) the bony cavity in a vertebrate skull that contains the eyeball (↓).

eyeball (*n*) in cephalopod molluscs and vertebrates the spherical structure that lies within the orbit (↑). It has three layers: the sclerotic (↓), choroid (↓) and retina (↓). The iris (↓) and lens (p.172) divide it into two chambers. The one lying at the front is filled with aqueous humour, the rear one with vitreous humour (p.172).

sclerotic, – coat, sclera (*n*) the tough, opaque, outer layer of the eyeball (↑) in cephalopod molluscs and vertebrates. It is fibrous or cartilaginous and protects the eyeball and helps maintain its shape. At the front of the eyeball, it is modified into the cornea (↓).

cornea (*n*) the transparent covering over the lens and iris at the front of the eyeball (↑) in cephalopod molluscs and vertebrates. It is a continuation of the sclerotic (↑). In terrestrial vertebrates it refracts light and helps the lens in the focusing of light on the retina (↓).

choroid, – coat, – membrane (*n*) the middle layer of the eyeball (↑) in cephalopod molluscs and vertebrates that lies within the sclerotic (↑) and external to the retina (↓). It contains blood vessels supplying the retina with nourishment and dark pigment that prevents internal reflection of light. The choroid forms the ciliary body and the iris (↓) at the front of the eye.

retina (*n*) the innermost, sensory layer of the eyeball (↑) in cephalopod molluscs and vertebrates; it does not extend over the front of the eyeball. It has a complex structure in which are found pigmented cells, rods, cones (p.173) and nerve cells whose fibres join the optic nerve. Light stimulates the rods and cones and impulses are conducted to the CNS in the optic nerve.

iris (*n*) a thin, circular, coloured ring of muscular tissue in the eyeball (↑) of cephalopod molluscs and vertebrates. Its outer rim is attached to the ciliary body. It lies over the lens and has a central opening, the pupil (p.172), which allows light to pass to the retina (↑). Except in some fish, the amount of light reaching the retina controls the size of the pupil by reflex action. Muscles in the iris can enlarge the pupil in dim light; another set of muscles contracts the pupil in strong light. The iris also assists in accommodation (p.174) for near objects.

pupil (*n*) the central opening in the iris (p.171) of cephalopod molluscs and vertebrates through which light passes to the retina. In most vertebrates, including man, it is round; in others, e.g. cat, it is slit-like in the day, but opens wide at night.

conjunctiva (*n*) protective transparent layer covering the cornea (p.171), becoming the opaque mucus-secreting inner lining of the eyelid (↓) of vertebrates.

eyelids (*n*) folds of skin and muscle, one lying above, the other below the eye of some vertebrates, e.g. man. They can be closed to protect the eye. In man, the lachrymal gland (↓) is found beneath the upper eyelid, and eyelashes (↓) are present on the edges.

eyelash (*n*) one of many hairs found on the edge of the eyelids (↑) in mammals. They help protect the cornea from damage by foreign bodies.

lachrymal gland, lachrimal gland, tear gland (*n*) a gland found on the eyeball in terrestrial vertebrates. In mammals it is beneath the upper eyelid (↑). It secretes **tears**, a salty sterile liquid which keeps the cornea moist and washes foreign bodies from the eyes. Tears drain through a **tear duct** at the inner corner of the eye into the nasal cavity.

aqueous humour (*n*) a clear, watery fluid that fills the space at the front of the vertebrate eyeball between the cornea and the lens (↓). It is secreted continuously by the ciliary body (↓). It is important in being the medium through which nutrients reach the lens and cornea. It also helps maintain the shape of the eyeball. *Compare vitreous humour* (↓).

vitreous humour (*n*) a clear, jelly-like substance which fills the cavity of the vertebrate eyeball behind the lens. It helps refract light and maintain the shape of the eyeball. *Compare aqueous humour* (↑).

lens (*n*) a transparent structure found just behind the pupil in the vertebrate eye. In terrestrial vertebrates it is biconvex in shape; in fish it is spherical. Suspensory ligaments (↓) attach it to the ciliary body (↓). It refracts light and thus helps focus an image on the retina. In mammals focusing is achieved by altering the shape of the lens; in fish the lens is moved back and forth, this is brought about by the muscles of the ciliary body (↓) *(see accommodation, p.174)*.

suspensory ligament (*n*) collagen fibres that attach the lens (↑) of the vertebrate eyeball to the ciliary body (↓).

lachrymal gland

position of lachrymal gland beneath upper eyelid

position of duct of lachrymal gland

tear ducts

pupil

eyelids   iris

eyelash   nasal cavity

nose

right mammalian eye showing lachrymal gland

rectus muscle

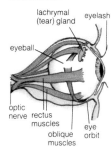

lachrymal (tear) gland

eyelash

eyeball

optic nerve rectus muscles

oblique muscles

eye orbit

rectus muscle of human eye

ciliary body (*n*) the muscular rim of the choroid at the edge of the cornea in the eyeball of cephalopod molluscs and vertebrates. The iris and suspensory ligaments (↑) are attached to it. It secretes aqueous humour (↑). **Ciliary muscle** in it can change the shape of the lens (↑) to bring about accommodation (p.174).

rectus muscle (*n*) one of four muscles that move the eyeball up and down or left and right.

oblique muscle (*n*) one of two muscles that enable the eye to roll and tilt in the socket.

blind spot, optic disc (*n*) the region of the vertebrate retina where the optic nerve enters the eyeball. Rods (↓) and cones (↓) are not found here so this region is not sensitive to light.

macula (*n*) an area of relatively acute vision in the retina of many vertebrates.

fovea, fovea centralis (*n*) a circular depression in the retina of some vertebrates including man. It contains many cones (↓) but no rods (↓). Light is most sharply focused on it and therefore it is the area of most acute vision.

yellow spot, macula lutea (*n*) an area surrounding the fovea (↑) of man and some primates; it contains yellow pigment.

rods (*n*) photoreceptors found in the retina of most vertebrates. They are rod-shaped and usually concentrated around the periphery of the retina. They contain visual purple (↓) and are responsible for vision in dim light. The only visual cells found in nocturnal animals are rods. *Compare cones* (↓).

visual purple, rhodopsin (*n*) a reddish-purple pigment found in the rods (↑) of the retina. Vitamin A is necessary for its formation. Light changes it to visual yellow and this reaction initiates nerve impulses. Lack of visual purple results in night blindness (p.128).

cones (*n*) photoreceptors found in the retina of most vertebrates; they are normally absent in nocturnal animals. The fovea (↑) and yellow spot (↑) contain many of them. Cones differ from rods (↑) in their shape and their lack of visual purple. They are concerned with the sharpness of vision, i.e. the discrimination of detail and with **colour vision**. There are three types of cone, each type sensitive only to red, blue or green light. Differential stimulation of the three types results in the sensation of colour.

accommodation (*n*) the changes in shape of the structures of the eye to give it refractory properties so that the image is focused on the retina. The cornea, aqueous humour, vitreous humour and lens all help refraction (↓) of light. Accommodation can be brought about by various means. In most birds and reptiles by ciliary muscles which squeeze the lens to alter its curvature; in fish and amphibians by movement of the lens backwards and forwards relative to the retina; in man and a few other mammals by altering the tension in suspensory ligaments and so altering the lens curvature; this is helped by constriction of the pupil when viewing near objects and dilation of it for distant objects.

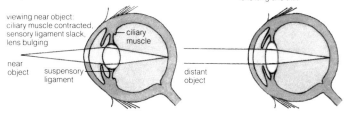

accommodation
accommodation of human eye

viewing distant object: ciliary muscle relaxed, suspensory ligament tensed lens long and thin

viewing near object: ciliary muscle contracted, sensory ligament slack, lens bulging

ciliary muscle

near object

suspensory ligament

distant object

refraction (*n*) the bending of light rays as they pass from one medium to another.
focus (*n*) the point at which light rays converge after refraction (↑) or reflection.
concave (*adj*) curved inwards, e.g. concave or **diverging lens** and **biconcave lens**. Compare convex (↓).
convex (*n*) curved outwards, e.g. convex or **converging lens** and **biconvex lens**. Compare concave (↑).
short sight, myopia (*n*) an eye defect, usually with reference to man where the eyeball is too long, hence only the images of near objects can be focused clearly; far objects are blurred as their images are focused in front of the retina. Myopia can be corrected by using diverging lenses.
long sight, hypermetropia (*n*) an eye defect, usually with reference to man where the eyeball is too short, hence only the images of distant objects can be focused clearly; near objects are blurred as their images are focused at a theoretical point behind the retina. Hypermetropia is corrected by using converging lenses.

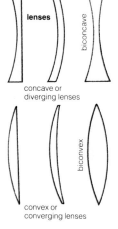

lenses

biconcave

concave or diverging lenses

biconvex

convex or converging lenses

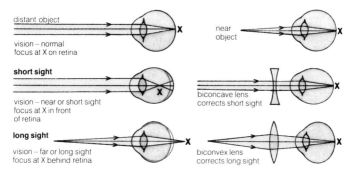

distant object

vision – normal
focus at X on retina

near object

short sight

vision – near or short sight
focus at X in front
of retina

biconcave lens
corrects short sight

long sight

vision – far or long sight
focus at X behind retina

biconvex lens
corrects long sight

astigmatism (*n*) an eye defect, usually with reference to
man, resulting from an irregularly shaped cornea. Rays
from one point do not have a single focus and so vision
is blurred. Astigmatism can be corrected with
cylindrical lenses.

lack of accommodation (*n*) an eye defect of elderly
people. Loss of elasticity in the lens and weakness of
the ciliary muscle leads to lack of accommodation (↑)
for near objects. This is similar to long sight (↑) and is
corrected similarly.

colour blindness (*n*) an inability to distinguish certain
colours. In man this is often the result of defective
cones.

conjunctivitis (*n*) inflammation of the conjunctiva
(p.172) of the eye.

trachoma (*n*) a viral disease of the eye in man. Hard
granules form on the lining of the eyelids and may result
in blindness.

cataract (*n*) a condition in man in which the lens of the
eye becomes opaque (↓). It is treated by surgical
removal of the lens and subsequent use of spectacles.

glaucoma (*n*) an eye disturbance in man caused by
faulty balance between production and drainage of
aqueous humour. It results in increased tension in the
eyeball and an increasing blurring of vision.

opaque (*adj*) describes a substance through which light
rays do not pass, i.e. it cannot be seen through.

translucent (*adj*) describes a substance through which
light rays pass although it is not perfectly transparent (↓).

transparent (*adj*) describes a substance through which
light rays pass, i.e. it can be seen through.

hearing (*n*) one of the senses in animals; it involves stimulation of auditory organs (↓) by sounds from the environment. It is poorly developed in most invertebrates and fish. It is well developed in insects, amphibians, reptiles, birds and mammals.

audible (*adj*) describes a sound which can be heard.

inaudible (*adj*) describes a sound not able to be heard.

auditory (*adj*) concerned with the sense of hearing (↑), e.g. the **auditory nerve** is the nerve of hearing.

acoustic (1) (*adj*) concerned with the sense of hearing (↑). (2) (*n*) the theory of sound.

otic (*adj*) concerned with the ear (↓).

deafness (*n*) inability to hear sounds.

auditory organ (*n*) the animal sense organ concerned with hearing (↑).

ear (*n*) the vertebrate auditory organ (↑). It not only detects sounds but also position relative to gravity and movement of the head relative to the body and is concerned therefore with both hearing and balance. The structure of the ear in different vertebrates varies. It is most developed in mammals where it is divided into the outer (↓), middle (↓) and inner ears (p.178).

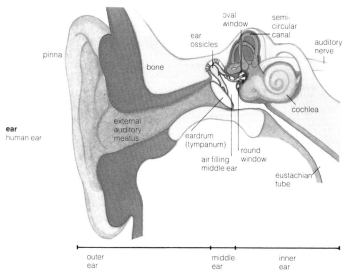

ear
human ear

outer ear (*n*) the parts of an ear (↑) that are external to the eardrum (↓). It is absent in amphibians and some reptiles where the eardrum is exposed on the skin surface. In other reptiles, birds and mammals it is a short tube, the external auditory meatus (↓). In mammals the pinna (↓) is found at its outer opening.

external auditory meatus (*n*) the passage in the outer ear (↑) of some reptiles, birds and mammals that leads from the outside to the eardrum (↓). It produces wax to keep the eardrum supple.

meatus (*n*) a channel or passage, e.g. external auditory meatus (↑).

pinna (*n*) a flap of skin and cartilage covering the external opening to the outer ear (↑) in mammals. Some mammals can move it to maximise the collection of sounds and help determine their direction.

eardrum, tympanic membrane (*n*) a thin, tough membrane stretched across the external opening of the middle ear (↓) of most amphibians, reptiles, birds and mammals. It vibrates when sound waves strike it; the vibrations are passed to the ear ossicles (↓).

tympanum (*n*) (1) the middle ear (↓) of vertebrates; (2) the eardrum (↑); (3) a vibratory membrane covering the auditory organ (↑) in various parts of an insect body which serves as an eardrum.

middle ear (*n*) the air-filled cavity between the eardrum (↑) and the oval window (p.179) in most tetrapod vertebrates. Sound is conducted across it to the inner ear (p.178) by the ear ossicles (↓). The Eustachian tube (↓) runs from it to the pharynx. The middle ear is absent in fish, newts, salamanders and snakes; here vibrations are transmitted to the inner ear through the body.

tympanic bone (*n*) the bone that supports the eardrum (↑) in amphibians, reptiles, birds and mammals.

Eustachian tube (*n*) a tube which runs from the middle ear (↑) to the pharynx in amphibians, reptiles, birds and mammals. It prevents distortion of the eardrum (↑) by allowing the air pressure between the middle ear and the atmosphere to equilibrate.

ear ossicle (*n*) a small bone in the middle ear (↑) of amphibians, reptiles, birds and mammals. The number present varies: reptiles have one ossicle; mammals have three, the malleus, incus and stapes (p.178). They transmit the vibrations of the eardrum to the inner ear (p.178) via the oval window (p.179).

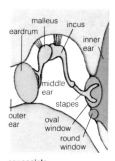

ear ossicle
ear ossicles of human ear

malleus (*n*) the outer ear ossicle (p.177) in mammals, lying between the eardrum and the incus (↓). Because of its shape it is sometimes called the **hammer**.

incus (*n*) the middle ear ossicle (p.177) in mammals, lying between the malleus (↑) and the stapes (↓). Because of its shape it is sometimes called the **anvil**.

stapes (*n*) the inner ear ossicle (p.177) in mammals lying between the incus (↑) and oval window (↓) of the inner ear. Because of its shape it is also called the **stirrup**.

inner ear (*n*) the innermost part of the vertebrate ear. It lies within the skull. It is connected to the middle ear (p.177) by the oval window (↓) in amphibians and reptiles and by the oval and round windows (↓) in other vertebrates. The **auditory nerve** runs from it to the brain. The inner ear consists of a membranous labyrinth (↓) filled with endolymph and contains the sensory receptors concerned with hearing and balance. The labyrinth is surrounded by perilymph (↓).

membranous labyrinth (*n*) closed system in the inner ear (↑) consisting of two sacs, the utriculus (↓) and sacculus (↓), connected by canals. Structures concerned with hearing and balance arise from the sacs. The membranous labyrinth is filled with endolymph (↓) and surrounded by another liquid, perilymph (↓).

perilymph (*n*) viscous liquid that surrounds structures of the membranous labyrinth (↑) of the inner ear and protects them from mechanical injury. It also conducts vibrations from the middle ear. Its composition is similar to cerebrospinal fluid. *Compare endolymph* (↓).

endolymph (*n*) a viscous fluid that fills the membranous labyrinth (↑) of the inner ear. Movement of the head causes it to move and stimulate sensory cells which play an important role in the sense of balance. It also conducts vibrations for hearing. *Compare perilymph* (↑).

sacculus, saccule (*n*) a small cavity of the membranous labyrinth (↑) of the inner ear. It is connected to the utriculus (↓) and filled with endolymph (↑). Areas of sensory cells, macula (↓), are found on the internal walls. In some vertebrates, e.g. birds and mammals, the cochlea (↓) arises from it.

utriculus, utricle (*n*) a cavity of the membranous labyrinth (↑) of the inner ear. It is connected to the smaller sacculus (↑) and is filled with endolymph (↑). Areas of sensory cells, macula (↓), are found on its internal walls In all vertebrates semicircular canals (↓) arise from it.

membranous labyrinth

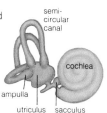

membranous labyrinth and cochlea of human ear

macula (*n.pl.*) areas containing sensory cells found in the sacculus (↑) and utriculus (↑) of the inner ear of vertebrates. The sensory cells are connected to the auditory nerve and have hairs which are surrounded by small granules of calcium carbonate called **otoliths**. These granules can press on the hairs provoking the sensory cells to send impulses along the auditory nerve to the brain so the animal is aware both of its position relative to gravity and of its rate of movement.

semicircular canals (*n.pl.*) tubes connected at both ends to the utriculus (↑) of the inner ear of vertebrates. Most have three set at right angles to each other. They contain endolymph. Each has a swelling at one end, an **ampulla**, containing sensory cells with protruding hairs embedded in a mucilaginous flap, the **cupula**. Attached to the hairs are small granules of calcium carbonate called otoliths (↑). These detect movement in the endolymph and initiate impulses along the auditory nerve to the brain so the animal is aware of any turning movement of the head, relative to the body.

cochlea (*n*) a projection of the sacculus (↑) in the inner ear of crocodiles, birds and mammals (in the latter it is coiled). It is connected to the middle ear by the oval window (↓) and by the round window (↓). It contains sensory cells lying on a membrane which vibrates when sound waves from the oval window reach the perilymph.

organ of corti (*n*) organ found in the cochlea (↑). Hair cells in it are stimulated by movement of endolymph to send impulses to the brain for interpretation of the pitch of a sound.

oval window, fenestra ovalis (*n*) a membrane-covered opening between the middle and inner ear of amphibians, reptiles, birds and mammals. Vibrations pass across it from the inner ear ossicle to the perilymph.

round window, fenestra rotunda (*n*) a membrane-covered opening between the middle and inner ear in highly developed ears, e.g. of birds and mammals. It allows the exit of pressure caused by the ear ossicle pressing on the oval window (↑).

lagena (*n*) projection from sacculus in reptiles and amphibians concerned with hearing.

echolocation (*n*) detection of position of objects by reflection from them of supersonic vibrations produced by an animal, e.g. bat, porpoise.

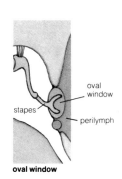

oval window
perilymph
stapes

oval window

endocrine system (*n*) a system of glands in animals which produce hormones (↓). The system interacts with the nervous system to control body functions.

endocrine gland, ductless gland (*n*) in animals, a gland that does not have a duct through which to secrete its product *(compare exocrine gland p.84)*. It produces one or more hormones (↓). The glands are found in some invertebrates and all vertebrates. Those in vertebrates include: the pituitary (p.167), thyroid (↓), parathyroid (p.182), thymus (p.144), adrenal (p.182) and the gonads (p.188).

hormone (*n*) (1) in animals an organic substance produced by endocrine glands (↑) which help co-ordinate body functions. Hormones may be cholesterol derivatives, e.g. steroid hormones; amino acid deriva-tives, e.g. thyroxin; polypeptides, e.g. insulin, TSH. They are secreted directly into the vascular system and carried by blood to the site(s) of their activity. Some affect the body generally, e.g. growth hormones (↓), others are more specific, e.g. gonadotropic hormones (↓) affect mainly sex organs. Hormones are required in small amounts. Their control is slower but usually longer lasting than that exercised by the nervous system (p.160) and thus the two methods of control are usually complementary. (2) in plants, an organic substance, e.g. an auxin (p.47), or growth hormone.

thyrotrop(h)ic hormone, thyroid stimulating hormone, TSH, thyrotrop(h)in (*n*) a peptide hormone (↑) secreted by the anterior lobe of the pituitary (p.167) of vertebrates. It stimulates growth of the thyroid gland and causes it to secrete thyroxin (↓).

adrenocorticotrop(h)ic hormone, ACTH (*n*) a polypeptide hormone (↑) secreted by the anterior lobe of the pituitary (p.167) of vertebrates. It controls the secretion of some of the hormones of the adrenal cortex, particularly the glucocorticoids.

gonadotrop(h)ic hormone, gonadotrop(h)in (*n*) one of several polypeptide hormones secreted by the anterior lobe of the pituitary (p.167) of vertebrates; concerned with maturation of, and production of, secretions by the gonads or sex glands, e.g. (a) **follicle-stimulating hor-mone, FSH**, in males causes sperm to mature and in females stimulates development of ovarian follicles and oestrogen production by the ovaries. (b) **luteinizing hormone, LH**, in females causes ovulation, secretion of

endocrine gland

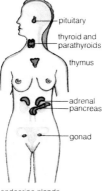

pituitary

thyroid and parathyroids

thymus

adrenal
pancreas

gonad

endocrine glands
of human female

oestrogens, stimulation of corpus luteum
formation. In males **LH** is called the **interstitial cell-
stimulating hormone, ICSH** and it stimulates
testosterone secretion.

**prolactin, lactogenic hormone, luteotrop(h)ic
hormone, LTH** (*n*) a polypeptide hormone secreted by
the anterior lobe of the pituitary. In lower vertebrates it
effects diverse activities, e.g. reproduction, growth,
osmoregulation. In birds it promotes secretion of 'milk'
from the crop glands. In female mammals it causes
development of breasts, milk production after childbirth
and secretion of progesterone by the corpus luteum.

growth hormone, somatotrop(h)ic hormone (*n*) a
polypeptide hormone secreted by the anterior lobe of
the pituitary of vertebrates which controls the growth of
the body. Excess of it causes **gigantism** where the long
bones increase greatly in length and the person is
abnormally tall, and **acromegaly** in which the bones of
the hands, feet and face enlarge and thicken. Deficiency
causes **dwarfism**; the person is abnormally small.

vasopressin, antidiuretic hormone, ADH (*n*) a
polypeptide hormone secreted in mammals by the
hypothalamus, but stored in the posterior lobe of the
pituitary (p.167). Acts mainly on collecting tubules of the
kidney causing reabsorption of water, thus decreasing
urine volume. Any condition causing dehydration
stimulates ADH release. Deficiency of ADH results in
diabetes insipidus (p.156). ADH also raises blood pressure by causing constriction of arterioles and capillaries.

oxytocin (*n*) a polypeptide hormone secreted by the
hypothalamus of birds and mammals but stored in the
posterior lobe of the pituitary. Its main actions in
mammals are expulsion of milk from lactating breasts
and contraction of the uterus.

thyroid gland, thyroid (*n*) an endocrine gland of
vertebrates. It has two lobes, one lying on each side of
the trachea. When stimulated by TSH (↑) from the
pituitary it secretes thyroxin (↓), an iodine-containing
hormone, which regulates metabolic rate. Lack of
iodine in the diet causes goitre (p.129).

thyroxin(e), thyroid hormone (*n*) an iodine-containing
amino acid hormone secreted by the vertebrate thyroid
(↑). It increases metabolism and is vital for normal
growth and development. Incorrect levels of it cause
disease, e.g. insufficient causes **cretinism** where

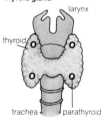

thyroid gland

larynx

thyroid

trachea — parathyroid

human thyroid and
parathyroid glands

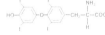

thyroxin
formula of thyroxin

physical and mental development ceases at an early
age, and **myxoedema** where the individual is obese
and mental activities are slow. Thyroxin also stimulates
moulting and metamorphosis of tadpoles.

parathyroid gland (*n*) an endocrine gland of tetrapod
vertebrates that lies in or near the thyroid (p.181). In
some species there are two, in man there are four lying
behind the thyroid and within the same capsule. It
secretes parathyroid hormone (↓).

parathyroid hormone, parathyrin, parathormone (*n*) a
polypeptide hormone secreted by the parathyroid
gland (↑). It controls the level of calcium in the blood.

adrenal gland, suprarenal gland (*n*) endocrine gland
in vertebrates. In man and other mammals there is one on
the upper, outer surface of each kidney. The outer part,
the **adrenal cortex**, secretes many hormones essential
to life. These fall into three classes: glucocorticoids (↓);
mineralocorticoids (↓); sex hormones (testosterone,
oestrogen, progresterone). Secretion is controlled by
ACTH. The inner part of the gland, the **adrenal
medulla**, secretes adrenalin (↓) and noradrenalin (↓)
when stimulated by the sympathetic nervous system.
These are not essential to life but are important in fight
and flight responses.

glucocorticoids (*n.pl.*) a group of steroid hormones
secreted by the adrenal cortex (↑) of vertebrates, e.g.
cortisol, corticosterone and cortisone (p.124). They
regulate metabolism of carbohydrates, fats and proteins.

mineral corticoids, mineralocorticoids (*n.pl.*) a group
of steroid hormones secreted by the adrenal cortex (↑)
of vertebrates. The main one is **aldosterone**. They
control the concentration of mineral ions in blood
plasma, lymph and tissue fluid.

adrenalin(e), epinephrine (*n*) the main hormone
secreted by the adrenal medulla (↑) under conditions of
stress, e.g. pain, fear, exercise, low blood sugar.
Adrenalin increases heartbeat, blood pressure and
blood sugar; dilates blood vessels of muscles, heart
and brain simultaneously causing contraction of those
in the skin and viscera; widens the pupil of the eye;
makes hair erect. Adrenalin is also secreted by some
invertebrates. *See noradrenalin* (↓).

noradrenalin(e), norepinephrine (*n*) a hormone
secreted by the adrenal medulla (↑). It is released under
the same conditions as adrenalin (↑) and has a similar

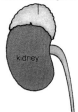

adrenal gland
human adrenal gland

kidney

CH_2OH

aldosterone
formula of aldosterone

formula of adrenalin
adrenalin

formula of noradrenalin
noradrenalin

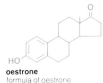

testosterone
formula of testosterone

oestrone
formula of oestrone

progesterone
formula of progesterone

function. Adrenalin and noradrenalin are secreted by the nerve endings of many sympathetic nerves and can act as neurotransmitters. Noradrenalin is also secreted by some invertebrates.

androgen (*n*) one of a group of steroid hormones that promote development of male sexual characteristics of vertebrates, e.g. testosterone (↓) and androsterone. Mainly produced by the testis but to a smaller extent by the ovaries and adrenal cortex (↑).

testosterone (*n*) an androgen (↑) produced by the testis of vertebrates. It is also an intermediate in the biosynthesis of oestrogens (↓).

oestrogen, estrogen (*n*) one of a group of steroid hormones of vertebrates, secreted mainly by the ovaries and placenta but also by the testis and adrenal cortex (↑), e.g. **oestradiol**, **oestriol** and **oestrone**. They have many metabolic effects, the chief one being stimulation in female mammals of the growth of reproductive organs and development of secondary sexual characteristics.

progesterone (*n*) a steroid hormone produced mainly by the corpus luteum in the ovary of mammals. It prepares the female reproductive organs for, and maintains the uterus during, pregnancy. It also promotes the development of secretory tissue in the mammary glands.

islets of Langerhans (*n*) groups of endocrine cells in the pancreas (p.107) of vertebrates: ∝ cells produce glucagon (↓); ß cells produce insulin (↓).

insulin (*n*) a hormone secreted by the ß cells of the islets of Langerhans (↑) which is involved in the control of blood sugar levels. Insulin increases the formation of glycogen by the liver and muscle and supresses its breakdown to glucose. It is antagonistic to glucagon (↓). Deficiency of it causes diabetes mellitus (p.156).

glucagon (*n*) a hormone secreted by the ∝ cells of the islets of Langerhans (↑) which is involved in the control of blood sugar levels. It stimulates the breakdown of glycogen to glucose. It is antagonistic to insulin (↑).

gastrointestinal hormones (*n.pl.*) hormones secreted by the epithelium of parts of the alimentary canal of vertebrates; e.g. (a) **secretin**, produced by the duodenum and jejunum in response to the presence of acidic food in the stomach, stimulates production of pancreatic juice and secretion of bile. (b) **gastrin**, secreted by the stomach, induces the secretion of gastric juice.

behaviour (*n*) the way an animal reacts to stimuli from the environment. It may be innate (↓) or as a result of learning (↓).

innate behaviour, instinct (*n*) an involuntary series of reactions by animals in response to certain stimuli. It is similar in all individuals of the same species. It is inherited and not dependent on reason (↓), experience (↓) or learning (↓); e.g. nest building by birds.

learning (*n*) the gaining of new knowledge, skill and behaviour patterns or their modification as a result of an animal's experience (↓).

conditioned reflex (*n*) the type of learning (↑) in which the instinctive reflex reactions of an animal in response to a given stimulus are altered as a result of the experience of the individual. When an animal keeps a conditioned reflex or habit (↓) it is said to **remember**; when it loses the response it is said to **forget.**

habit (*n*) an often used behaviour pattern (↑) of an animal that has been learnt by continuous repetition.

experience (*n*) knowledge gained by an animal as a result of encountering certain events previously.

memory (*n*) the ability to store in the brain information learnt as a result of the animal's experience (↑).

will (*n*) the ability found in higher animals to decide which way to act in response to a particular set of stimuli. Such animals can also initiate actions in the absence of external stimuli and independently of habit.

intelligence (*n*) the mental skill of an animal that enables it to make correct responses to new circumstances.

mental (*adj*) concerned with the mind, i.e. the processes that take place in the brain of an animal.

reason (*n*) ability of the brain of higher animals to decide the possible results of various responses to a particular set of circumstances. The animal is then able to select the most appropriate response(s).

emotion (*n*) a strong reaction of the mind, usually of man, to stimuli from the environment. Emotions include fear, anger, love, hate.

drive (*n*) the motivation of an animal to carry out a set of actions, e.g. hunger drive makes an animal seek food.

irritability (*n*) the ability of a cell, tissue or organism to respond to changes in the environment. In plants it is shown by changes in direction of growth in response to stimuli, e.g. light, water, etc., and in animals in activities controlled by the nervous system.

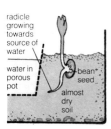

radicle growing towards source of water

water in porous pot

bean seed

almost dry soil

irritability
irritability in plants

growth

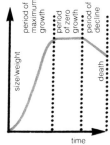

growth curve of
multicellular organisms
– man

growth (*n*) a gradual increase in size and/or weight of an organism or part of it as development occurs. It usually involves both increase in cell size and cell numbers (i.e. cell division). Growth rate varies during the life of an organism, often showing a **growth curve** (*see diagram*).

movement (*n*) a change in position of a plant or animal. Plant movements are usually growth responses (i.e. tropisms p.46). Animal movements involve either (a) altering relative position of body parts, e.g. waving the arms or (b) transfer of the animal from place to place.

locomotion (*n*) movement (↑) from place to place, e.g. walking, running. This is shown by animals and is brought about by movement of body parts by muscles.

mobile (*adj*) describes animals with locomotion (↑) that can move from place to place. *Compare motile* p.28.

sedentary (*adj*) describes animals lacking locomotion (↑) and hence unable to move from place to place.

biological clock (*n*) system in plants and animals which controls circadian rhythms (↓).

circadian rhythm, diurnal rhythm (*n*) the rhythm (↓) of changes occurring in living organisms. They are normally synchronised with the daily light/dark cycle and are largely independent of other external factors in the environment, e.g. leaf movements and alternation of sleep and activity in animals both have 24-hr cycles.

rhythm (*n*) a regular repetition of a series of events, e.g. rhythm of heartbeat; changes in the circadian rhythm (↑).

nocturnal (*adj*) occurring at night, e.g. nocturnal animals are active during the hours of darkness.

diurnal (*adj*) describes an event (1) occurring daily, e.g. that with circadian rhythm (↑) (2) occurring during the daytime, e.g. an animal that is active during daylight.

hibernation (*n*) dormancy in animals that occurs during winter. It is a state of deep sleep when metabolic processes are slow and body temperature falls.

aestivation (*n*) (1) in animals (e.g. lung fish), dormancy during summer or dry season. (2) in plants, the way in which flower parts are folded in a flower bud.

migration (*n*) a movement, usually seasonal, of populations of animals from one region to another; it is particularly common in birds. Migration to a warmer climate occurs as temperatures fall, hours of daylight shorten and food is less plentiful; the animals return when these conditions are reversed. In some species it is an alternative to hibernation (↑) for surviving the winter.

reproduction (*n*) the process by which plants and animals produce new individuals. There are two types: asexual reproduction (↓) which involves one parent and sexual reproduction (↓) which involves two parents.

asexual reproduction (*n*) a type of reproduction (↑) in which an organism produces new individuals without the formation of gametes. It is common in plants where it usually involves vegetative propagation (p.77) or spore production. It is rare in animals except amongst protozoa and other micro-organisms where it involves processes such as budding (in yeast) and binary fission. The offspring in asexual reproduction are genetically identical to the parent and to each other. *Compare sexual reproduction* (↓).

sexual reproduction (*n*) a type of reproduction (↑) involving the fusion of gametes in the production of new individuals. The process usually involves the fusion of a small, motile gamete (↓) from a male individual with a large, non-motile gamete from a female. However, in many plants and some lower animals, the two types of gamete are produced by one hermaphrodite (↓) organism. Because segregation and recombination (p.207) occur during gamete formation, sexually reproduced offspring are not identical with either the parents or each other. *Compare asexual reproduction* (↑).

propagation (*n*) reproduction (↑) in plants, either by an asexual process (↑), e.g. vegetative propagation (p.77), or by sexual reproduction (↑), e.g. seed production.

proliferation (*n*) a rapid increase in the numbers of cells or organisms by repeated reproduction (↑).

prolific (*adj*) describes cells, plants or animals that are reproducing rapidly.

parthenogenesis (*n*) reproduction (↑) involving the formation of an offspring from an ovum which has not been fertilized by a male gamete. The offspring are genetically identical to the parent. Pathenogenesis is found in some animals, e.g. aphids, and in some plants.

parthenocarpy (*n*) formation of fruits without fertilization. Fruits are seedless, e.g. seedless oranges.

unisexual (*adj*) describes (1) animals that produce male or female gametes, but not both; (2) flowering plants where the stamens and carpels are on separate flowers. In monoecious plants the two types of flower are on the same plant; in dioecious ones they are on separate plants. *Compare bisexual* (↓).

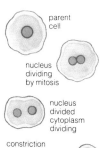

parent cell

nucleus dividing by mitosis

nucleus divided cytoplasm dividing

constriction of cytoplasm

two daughter cells

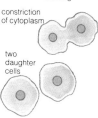

asexual reproduction
asexual reproduction by binary fission in *Amoeba*

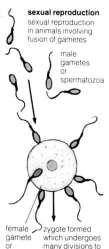

sexual reproduction
sexual reproduction in animals involving fusion of gametes

male gametes or spermatozoa

female gamete or ovum

zygote formed which undergoes many divisions to form new individual

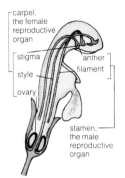

bisexual
sexual reproduction
in plants – bisexual flower
of dead nettle

Labels on figure: carpel, the female reproductive organ; stigma; style; ovary; anther; filament; stamen, the male reproductive organ

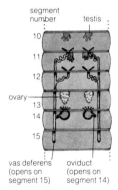

hermaphrodite
testis and ovary of
hermaphrodite earthworm

Labels on figure: segment number; testis; 10; 11; 12; 13; 14; 15; ovary; vas deferens (opens on segment 15); oviduct (opens on segment 14)

bisexual (*adj*) describes (1) animals that produce both
male and female gametes; (2) flowering plants that
have both stamens and carpels on the same flower.
Compare unisexual (↑).

hermaphrodite (*n*) a flowering plant or animal with both
male and female sexual structures on the same
individual.

neuter (*n*) an individual that has neither male nor female
reproductive organs.

gamete (*n*) a specialised cell in plants (see p.66) and
animals involved in sexual reproduction (↑). It has the
potential to fuse with another gamete at fertilization (↓)
to form a zygote (↓) which develops into a new
individual. Gametes are usually haploid (p.204). There
are usually two types, in animals the male gamete or
spermatozoon and the female gamete or ovum which
differ in size and motility.

gametocyte (*n*) a cell from which gametes (↑) are formed
by meiosis.

fertilization (*n*) the fusion of male and female gametes
during sexual reproduction in plants and animals. It
results in a single cell, the zygote (↓), from which the
new individual develops. **External fertilization** occurs
in many aquatic animals; fusion of gametes takes place
outside the body of the female. **Internal fertilization**
occurs in most terrestrial animals; union of gametes
takes place inside the body of the female.

insemination (*n*) the introduction, in animals, of
spermatozoa into the female, usually by artificial
means, i.e. **artificial insemination**. This is frequently
used in animal husbandry.

coition, coitus, copulation (*n*) the sexual act between a
male and a female animal. During it spermatozoa are
deposited by the male in the body of the female in order
to fertilize (↑) the ova.

conjugation (*n*) (1) fusion of two gametes in sexual
reproduction (↑), particularly where they are of
similar structure, i.e. isogametes. (2) union of two
individuals in sexual reproduction of most ciliates,
e.g. *Paramecium* and in some algae. (3) union of two
bacteria during which genetic information is
transferred.

zygote (*n*) diploid cell formed in plants (p.32) and
animals by the fusion of two haploid gametes (↑). It
develops into the embryo.

gonad (*n*) in animals, a sexual organ or reproductive
gland that produces gametes (p.187). In males, these
are the testis, in females, the ovaries and in a few
hermaphrodites, the ovotestis (↓). The gonads of some
animals also produce hormones.

ovotestis (*n*) a gonad (↑) found in some hermaphrodite
animals, e.g. snail. It functions as both ovary and testis,
i.e. it produces both eggs and spermatozoa.

genital organs, genitalia, genitals (*n.pl.*) in animals, the
organs that are involved in sexual reproduction. They
include especially the gonads (↑) and their associated
structures. The term is particularly used for the external
sex organs, e.g. the penis (↓) and scrotum (↓) in males;
the labia and clitoris (p.192) in females.

testis, testicle (*n*) the gonad (↑) found in male animals
that produces spermatozoa (p.190) and in some
animals male hormones also (see androgens p.183). In
most vertebrates it is composed of seminiferous tubules
(↓) and hormone-producing cells. The testes of
mammals lie within the scrotum (↓).

seminiferous tubules (*n*) coiled tubes found in the testis
(↑) of vertebrates. Spermatogenesis (p.190) occurs in
them.

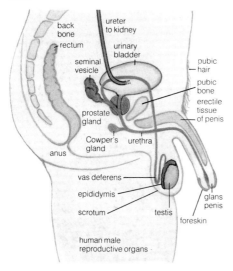

human male
reproductive organs

germinal epithelium (*n*) epithelium in the testis (↑) and ovary (p.190) of vertebrates which produces respectively spermatogonia and oogonia (p.190).

vas efferens (*n*) one of many tubes for the passage of spermatozoa (p.190) from a vertebrate testis (↑) to the epididymis (↓).

epididymis (*n*) a long convoluted tube found in each vertebrate testis (↑). It receives immature spermatozoa (p.190) from the seminiferous tubules via the vas efferens (↑). The spermatozoa mature in the epididymis before they pass to the vas deferens (↓).

vas deferens, sperm duct (*n*) a tube arising from the epididymis (↑) of male reptiles, birds, mammals and some amphibians. It carries spermatozoa (p.190) towards the exterior during copulation.

scrotum, scrotal sac (*n*) a pouch of skin in mammals that contains the testes (↑). It lies outside the abdominal cavity and so keeps the testes at an optimum temperature for production of spermatozoa (p.190).

castration (*n*) removal of (1) the testes (↑) from an animal, (2) the androecium from a flower.

penis (*n*) the male sexual organ of some reptiles, a few birds and all mammals. It contains erectile tissue (↓), which, when filled with blood, causes the penis to enlarge and become firm. This enables it to penetrate the female during copulation and thus allows the passage of spermatozoa from the male to the female. In mammals the distal end is covered by the foreskin (↓).

foreskin, prepuce (*n*) loose fold of skin at the end of the penis (↑) in male mammals. It covers the bulging portion, the **glans penis**.

circumcision (*n*) surgical removal of all or part of the foreskin (↑).

erectile tissue (*n*) spongy, vascular tissue which becomes swollen and firm when blood is pumped into it. It is found, for example, in the penis (↑).

seminal vesicle (*n*) (1) a sac found in the male of lower vertebrates and some invertebrates; it stores sperm. (2) a gland in the male of higher vertebrates and mammals; it lies behind the urinary bladder. It secretes into the vas deferens (↑) some of the fluid found in semen (p.190).

prostate gland (*n*) a gland in male mammals. It lies below the urinary bladder and surrounds part of the urethra. It secretes some of the substances that make up semen (p.190).

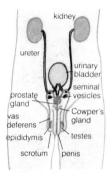

position of male reproductive organs in humans

Cowper's gland (*n*) one of two glands found below the prostate (p.189) in male mammals. It secretes some of the substances that make up semen (↓).

semen (*n*) the spermatozoa (↓) plus fluid secreted by the various glands associated with the male reproductive system. In mammals these are the seminal vesicle (p.189), prostate (p.189) and Cowper's glands (↑). Semen is ejaculated from the penis (p.189) during copulation.

seminal fluid (*n*) fluid secreted by the glands associated with the male reproductive system. It activates and nourishes spermatozoa (↓) and neutralizes the acidity of the vagina.

spermatozoon (*pl.spermatozoa*), **sperm** (*n*) small, motile, haploid gamete produced in the testis of male animals. It usually has a flagellum and consists of a large nucleus and little cytoplasm. It fertilizes the female gamete in the process of sexual reproduction.

spermatogonium (*n*) a cell in the testis of male animals that divides by mitosis (p.201) to produce spermatocytes (↓).

spermatocyte (*n*) a cell in an animal testis produced from a spermatogonium (↑). It undergoes division by meiosis (p.201) to form spermatozoa (↑): the **primary spermatocyte** divides to produce two **secondary spermatocytes**; each of these divides to form two **spermatids** that eventually develop into spermatozoa.

spermatogenesis (*n*) the formation of spermatozoa (↑).

ovary (*n*) the gonad in female animals; it produces ova. In vertebrates there is a pair of ovaries which produce hormones as well as ova in a series of cycles (see oestrous cycle p.192); this is controlled by gonadotrophins secreted by the pituitary. In mammals the surface layer of the ovary, the germinal epithelium (p.189), surrounds many Graafian follicles (↓). Each follicle surrounds one oogonium (↓). A few of these follicles mature and burst releasing oocytes (↓) at ovulation.

ovum (*pl.ova*), **egg cell** (*n*) a mature, unfertilized female gamete of animals. It is a large, non-motile cell containing abundant cytoplasm, yolk and a haploid nucleus. It is normally produced in the ovary (↑).

oogonium (*n*) (*pl. oogonia*), a cell in an animal ovary (↑) that divides by mitosis (p.201) to produce oocytes (↓).

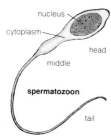

nucleus
cytoplasm
head
middle

spermatozoon

tail

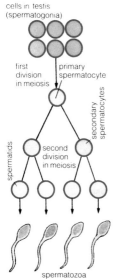

cells in testis (spermatogonia)

first division in meiosis
primary spermatocyte

secondary spermatocytes

spermatids
second division in meiosis

spermatozoa

spermatogenesis

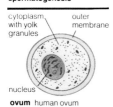

cytoplasm with yolk granules
outer membrane

nucleus

ovum human ovum

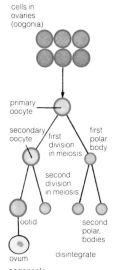

cells in
ovaries
(oogonia)

primary
oocyte

secondary
oocyte | first
division
in meiosis | first
polar
body

second
division
in meiosis

ootid | second
polar
bodies

ovum | disintegrate

oogenesis

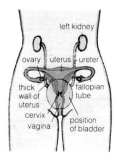

left kidney

ovary | uterus | ureter

thick
wall of
uterus | fallopian
tube

cervix
vagina | position
of bladder

position of female
reproductive organs in
humans

oocyte (*n*) a cell in an animal ovary (↑) produced from an oogonium (↑). It undergoes division by meiosis (p.201) to form an ovum (↑): the **primary oocyte** divides to form two cells, a **polar body** with little cytoplasm and a **secondary oocyte** with much cytoplasm; the secondary oocyte divides to form a further polar body and an **ootid**, which develops into the ovum (*see diagram*). In some species fertilization occurs at one of the two oocyte stages, i.e. without the formation of an ovum.

ovarian follicle (*n*) a sac of cells that invests and provides nourishment for the developing oocyte (↑) in vertebrates. It also secretes female sex hormones. In mammals it is called the **Graafian follicle**; this differs from an ovarian follicle in that it has a cavity and in most cases it forms the corpus luteum (↓) after ovulation.

corpus luteum (*n*) a temporary body formed after ovulation in the Graafian follicle of a mammalian ovary; it secretes progesterone. Its formation is initiated by an increase in the secretion of luteinizing hormone (p.180) by the pituitary. If fertilization of the ovum does not occur the corpus luteum degenerates; if fertilization occurs it persists for part or all of the period of pregnancy.

oogenesis, ovogenesis (*n*) the formation of ova (↑) usually in an ovary.

oogamy (*n*) sexual reproduction involving the fertilization of a large, non-motile female gamete or ovum by a small, motile male gamete. It occurs in animals and some plants.

ovulation (*n*) discharge from an ovarian follicle (↑) (Graafian follicle (↑) in mammals) of a ripe ovum or oocyte (↑). It passes into the oviduct (↓) or the fallopian tube (↓). Ovulation is initiated by hormones from the pituitary gland.

oviduct (*n*) tube in female animals which carries ova (↑) from the ovaries to the exterior. *Compare fallopian tube* (↓).

ovisac (*n*) a sac found in amphibians in which ova (↑) are held before discharge.

fallopian tube (*n*) tube in female mammals that carries ova (↑) from the ovary (↑) to the uterus. Spermatozoa (↑) may pass from the uterus to the upper part of the fallopian tube where fertilization of the ovum often occurs. *Compare oviduct* (↑).

uterus, womb (*n*) an organ of the reproductive system in nearly all female mammals. Usually two are present, one connected to each fallopian tube (p.191); in humans, however, there is only one. The uterus is connected via the cervix (↓) to the vagina (↓). During pregnancy it contains the developing embryo or foetus and provides it with nourishment via its glandular lining; *see endometrium* (↓). Its muscular walls increase in thickness during pregnancy and contract at birth to expel the foetus (p.195).

oestrus cycle (*n*) reproductive cycle associated with ovulation in sexually mature female mammals. Its length varies in different species from 4 to 6 days. The cycle is controlled by hormones secreted by the pituitary. At a particular phase of the cycle the female is receptive to copulation with the male. In most animals, each cycle is followed immediately by another one, until pregnancy occurs. However, in animals that show a definite breeding season, only one cycle will occur per season.

endometrium (*n*) mucous glandular membrane lining the uterus (↑) of mammals. Progesterone and oestrogen control its thickness and structure which vary during the oestrus cycle (↑). If the ovum is not fertilized the endometrium is destroyed and menstruation follows.

cervix (*n*) neck of the uterus (↑) which protrudes into the vagina (↓). Glands in it supply the vagina with mucus.

vagina (*n*) duct found in most female mammals that connects the uterus/uteri (↑) with the exterior by a short vestibule (↓). It receives the penis during copulation and acts as a birth canal for the foetus (p.195).

vestibule (*n*) small chamber in most female mammals into which the urethra (p.155) and vagina (↑) open. In humans it lies between the labia minor (↓).

labia (*n*) in female humans the lips that enclose the openings of the vagina (↑) and urethra. There are two pairs, the outer **labia major** and the inner **labia minor**.

hymen (*n*) thin membrane partially closing the vagina in a virgin woman. It is ruptured when coitus first occurs.

clitoris (*n*) small structure found near the opening of the urethra (p.155) in mammals. It consists of sensitive, erectile tissue. It is homologous with the male penis.

vulva (*n*) external genital organs of female humans; it includes the labia (↑), clitoris (↑) and vestibule (↑).

conception (*n*) the fertilization of an ovum in humans. The woman is then pregnant (↓). **conceive** (*v*).

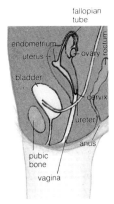

female reproductive organs of humans (side view)

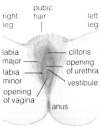

external genitalia of human female

pregnancy (*n*) the condition of a female mammal when an embryo is developing in the uterus (↑). The female is described as being **pregnant** (*adj*).

gestation (*n*) the developmental processes of a mammalian embryo that occur while it is being carried in the uterus (↑). The period between conception (↑) and birth is the **gestation period**.

contraception (*n*) any method of preventing pregnancy.

menstrual cycle

time and sequence of events in the menstrual cycle (times are approximate)

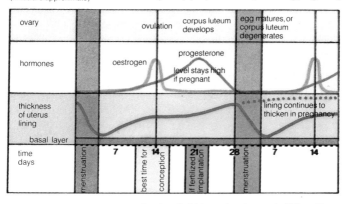

ovary			ovulation	corpus luteum develops	egg matures, or corpus luteum degenerates	
hormones		oestrogen		progesterone level stays high if pregnant		
thickness of uterus lining basal layer					lining continues to thicken in pregnancy	
time days	menstruation	7	best time for conception 14	if fertilized implantation 21	menstruation 28	7 14

menstrual cycle (*n*) type of oestrus cycle (↑) found in higher primates (e.g. chimpanzee, man) in which the uterus lining breaks down suddenly *(see diagram)* producing bleeding at menstruation (↓). The animal is sexually receptive to the male all through the cycle.

menstruation (*n*) periodic loss of blood from the uterus in the menstrual cycle (↑). In humans it occurs about every twenty-eight days from puberty to the menopause (↓).

'heat', (on heat) (*n*) period when female animal is sexually receptive to the male and ready for mating.

menopause (*n*) in female humans, the time when the menstrual cycle (↑) stops and reproduction is no longer possible.

egg membrane (*n*) protective and sometimes nutritive membrane surrounding a fertilized animal ovum (egg), e.g. (a) the **vitelline membrane** or **fertilization membrane,** which surrounds most ova, is secreted by the ovum to prevent the entry of more spermatozoa; (b) the outer shell of a bird's egg or jelly of an amphibian egg, both secreted by the oviduct.

embryo (*n*) in animals, the structure formed from the zygote (p.187) which develops into a new individual. In viviparous animals it lies within the uterus, in oviparous animals it is within the egg membrane (p.193). The embryo obtains nourishment from the parent or from the egg. It passes through a series of stages (**cleavage, blastula, gastrula**) involving repeated mitotic divisions. The outer layer of cells thus formed is the **ectoderm**, the middle layer the **mesoderm** and the inner layer the **endoderm**. Each differentiate into particular tissues and organs of the new individual. The embryonic period ends with birth or hatching of the egg.

embryonic (*adj*) concerned with (1) an embryo (↑) (2) any tissue or organ in its initial stages of development.

embryonic, extra-embryonic or **foetal membranes** (*n*) membranes that surround a vertebrate embryo during its development. They are the yolk sac (↓), allantois (↓), amnion (↓) and chorion (↓) and are involved in respiration, nutrition, excretion and protection of the embryo/foetus.

yolk sac (*n*) an embryonic membrane (↑) found in some vertebrates. In most fish, reptiles and birds but not mammals it contains yolk (↓).

yolk (*n*) a store of food (mainly protein and fat) found in the eggs of many animals for use by the developing embryo. The amount and position of it is variable, e.g. in birds it is yellow and found in the middle of the egg surrounded by the egg white. **Microlecithal** eggs have little yolk, **mesolecithal** moderate amounts and **macrolecithal** (or **megalecithal**) have relatively large amounts.

allantois (*n*) an embryonic membrane (↑) of reptiles, birds and mammals. In reptiles and birds it stores excretory waste products and acts as a respiratory surface. In mammals its most important role is in the nourishment of the embryo/foetus.

amnion (*n*) an embryonic membrane of reptiles, birds and mammals which encloses the amniotic fluid (↓).

chorion (*n*) the outer embryonic membrane (↑) in reptiles, birds and mammals. In mammals the placenta (↓) is formed from it.

amniotic fluid (*n*) liquid enclosed by the amnion (↑). It provides a fluid environment for the embryo/foetus and acts as a cushion to protect it from mechanical damage.

embryo

• embryo at 14 days

embryo at 28 days

foetus at 9 weeks

foetus at 15 weeks

relative size of developing human foetus during gestation

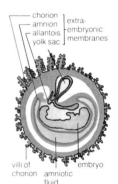

chorion
amnion   extra-embryonic
allantois   membranes
yolk sac

villi of chorion   embryo
amniotic fluid

extra-embryonic membranes
extra-embryonic membranes of human embryo

egg

egg of birds and reptiles

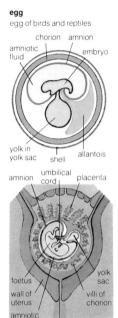

foetus

developing human foetus

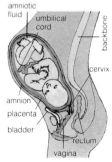

parturition human foetus just before parturition (birth)

egg (n) (1) structure, often oval, laid by birds, reptiles and amphibians. It consists of an ovum plus any materials needed for its development (e.g. yolk, albumen) and its protection (e.g. membranes, shell (↓)). (2) an ovum or female gamete, i.e. an egg cell of any animal.

shell (n) hard, thin outer covering of the egg (↑) in terrestrial oviparous animals. It allows the passage of gases, but not liquids.

implantation, nidation (n) attachment of a mammalian embryo to the lining of the uterus. This occurs after fertilization and before the formation of the placenta (↓).

placenta (n) a vascular structure found in the pregnant mammals. It is formed from the endometrium (p.192) of the mother and the chorion (↑) of the embryo. In it, because of placental villi, there is a close association between embryonic and maternal blood systems; the two, however, remain entirely separate. Oxygen and nutrients pass to the embryo; waste metabolic products, e.g. urea, carbon dioxide, pass from it to the mother. It also produces hormones. At birth the placenta is discharged after the foetus (↓) as the afterbirth.

foetus (n) a mammalian embryo after it has developed recognisable features. In man this is after 4 to 6 weeks.

umbilical cord (n) cord connecting the embryo/foetus to the placenta. It is either broken at birth or is cut. Passing through it are the two **umbilical arteries** and one **umbilical vein**.

parturition, birth (n) the expulsion of a mammalian foetus at the end of pregnancy. The amnion and chorion rupture, the uterus walls contract periodically and the foetus passes down the vagina, normally head first.

afterbirth (n) the placenta and membranes expelled after the foetus from the uterus at parturition (↑).

stillbirth (n) birth of a dead foetus.

caul (n) covering over a baby's head at birth. It is the amnion which is not completely ruptured.

viviparous (adj) describes an animal in which embryonic development occurs within the mother and from which it derives nutrients via a placenta, e.g. all placenta mammals including man. Compare oviparous (↓), ovoviviparous (p.196).

oviparous (n) describes an animal that lays fertilized eggs containing undeveloped embryos and a large amount of yolk, e.g. birds. Compare viviparous (↑), ovoviviparous (p.196).

ovoviviparous (*adj*) describes an animal in which embryonic development occurs within the mother but utilizes mainly food stored within the egg, e.g. many insects, fish and snakes. *Compare viviparous, oviparous* (p.195).

mammary glands, breasts, mammae (*n*) glands found on the ventral surface of female mammals. Their growth and activity is controlled by hormones secreted by the gonads. Mammary glands markedly increase in size after sexual maturity. They consist of clusters of gland cells that produce milk (see lactation (↓)). Milk emerges in most mammals from a teat (↓) or a nipple (↓).

udder (*n*) mammary glands (↑) of herbivores, e.g. cow, goat.

teat (*n*) small swelling on a mammalian mammary gland (↑) through which young suck milk. *Compare nipple* (↓).

nipple (*n*) the teat (↑) of a mammary gland, especially in humans.

lactation (*n*) production of milk by a mammary gland (↑) after parturition, being stimulated by several hormones. Sucking at the nipple helps maintain lactation.

colostrum (*n*) the first milk of mammals after parturition. It is very rich in proteins and in some mammals, including man, it contains antibodies.

suckle (*v*) to feed a young mammal with milk from the teat/nipple (↑) of mammary glands (↑).

secondary sexual characteristics (*n*) the differences between male and female animals that appear only after sexual maturity (i.e. excluding the gonads and related structures which are the **primary sexual characteristics**). These include in the male deepening of the voice and development of pubic and facial hair; in the female swelling of breasts, broadening of thighs, development of pubic hair and start of menstruation.

sex (*n*) division of organisms into males and females. This is rare in plants but common in animals where it is determined by the nature of the sex chromosomes (↓).

sex chromosomes (*n*) the pair of chromosomes which is associated with the sex of an animal. Normally there are two types designated **X** and **Y**. The latter is usually smaller. In most animals the female has two X chromosomes and the male one X and one Y.

heterogametic animals have dissimilar sex chromosomes (i.e. XY = male); **homogametic** animals have similar ones (i.e. XX = female).

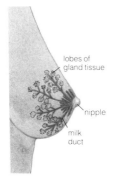

mammary gland
human breast
(side view)

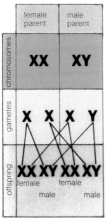

sex determination

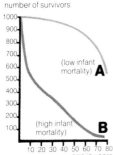

survival

number of survivors

(low infant mortality) **A**

(high infant mortality) **B**

age in years

number of survivors relative to age in countries A and B for 1980

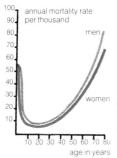

annual mortality rate per thousand

men

women

age in years

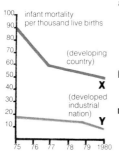

infant mortality per thousand live births

(developing country)

X

(developed industrial nation)

Y

75 76 77 78 79 1980

life span (*n*) time between the beginning and end of an organism's life. It varies from a few hours for adult stages of some insects through several years for some mammals to hundreds of years for some trees, e.g. oak.

life history (*n*) the changes of an organism from formation or birth, through maturity, to death.

age (*n*) (1) the number of years an organism has existed; (2) in geology a very long period in the history of the earth; (3) a period of time in the evolution of man, e.g. the iron age during which iron was discovered and used to make tools, etc.

birth rate (*n*) in man, the number of births relative to the total number of individuals; often expressed as the number of live births in one year, per 1000 of the total population.

sex ratio (*n*) the relative number of males to females in a population.

survival, survival rate (*n*) the number of individuals in a particular age-group remaining alive in a unit of time for every 1000 initial population of that age.

mortality, mortality rate (*n*) the number of deaths in a particular age-group in a unit of time for every 1000 initial population of that age.

infant mortality (*n*) in man, the proportion of deaths during the first year of life relative to the number of live births in thousands.

death rate (*n*) the number of deaths relative to the total population in thousands.

heredity (*n*) the transmission of characteristics (p.198) from one generation (p.198) to successive generations of living organisms. This is brought about by genetic information carried in the genes (p.205).

acquired characteristic (*n*) a characteristic (↓) developed by a living organism during its lifetime in response to environmental factors, e.g. muscular development in response to exercise. Inheritance of acquired characteristics first proposed by Lamarck (hence called **Lamarckism**). *Compare hereditary characteristic* (↓).

hereditary characteristic (*n*) an inherited characteristic (p.198), transmitted from one generation to the next, e.g. colour of eyes. *Compare acquired characteristic* (↑).

nature and nurture (*n*) the genetic make-up (nature) and the environment (nurture) of an organism. Both interact and affect the characteristics (p.198) of the individual.

characteristic, trait (*n*) a distinctive feature or quality of a living organism, e.g. blue eyes, wrinkled seeds. Combinations of characteristics are used in classification to distinguish individuals, varieties and species, etc. In different varieties most of the traits are similar and only a few vary; in different species fewer traits are similar and more vary.

inherit (*v*) to receive genes from the parent or parents as a result of reproduction.

inheritance (*n*) the combination of genes derived from the parent or parents as a result of reproduction.

pedigree (*n*) a record, often in the form of a table or chart, showing the ancestors (↓) of an animal.

generation (*n*) all the offspring from a set of parents, i.e. one level in a pedigree (↑).

ancestors (*n.pl.*) those from whom a particular individual has descended, i.e. in man, parents, grandparents, etc.

parent (*n*) a living organism that has produced a new individual. Usually this involves sexual reproduction. In animals, the female parent is the **mother**, the male parent is the **father**.

progeny (*n*) individuals descended over more than one generation from a particular group of parents (↑).

offspring (*n*) the first generation of progeny (↑) produced by a particular set of parents (↑).

siblings (*n.pl.*) offspring (↑) produced by the same male and female, i.e. in man, brothers and sisters.

descendants (*n.pl.*) the progeny (↑) derived from a particular set of parents (↑), e.g. in man, children, grandchildren, great grandchildren.

variation (*n*) differences in characteristics (↑) between individual plants or animals of the same species. These may be due to differences in genetic make-up and/or environmental factors (see genotype and phenotype p.205). Groups of individuals with the same variation or combination of variations are called **variants**.

clone (*n*) a group of genetically identical individuals produced from a single parent by non-sexual reproduction, e.g. in plants by vegetative propagation; in animals by parthenogenesis.

pure strain, pure line (*n*) a group of organisms of the same species that are homozygous for the gene or genes under consideration (strictly all genes). It is produced either by intensive inbreeding of an initial population, or by employing self-fertilization.

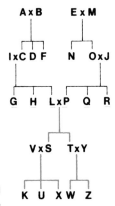

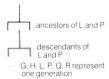

pedigree
pedigree of individuals
L and P and their descendants

hybrid (*n*) organism produced by sexual reproduction from parents that are not genetically identical. The hybrid produced is heterozygous for many of its genes. Plant hybrids are often larger and healthier than the parents; this is **hybrid vigour** or **heterosis**. Use of F_1 hybrids of seeds often gives increased crop yields.

breeding (*n*) production of progeny in plants and animals, particularly by the mating of selected parents with different, desirable characteristics in order to produce offspring with the desired combination of characteristics, e.g. new varieties of maize and rice which combine high yield with disease resistance.

inbreeding (*n*) breeding (↑) by mating of closely related and hence genetically similar individuals (*contrast hybrid* (↑)). It has disadvantages over outbreeding (↓) because it can result in the appearance of harmful recessive characteristics. *Contrast outbreeding* (↓).

outbreeding (*n*) breeding (↑) by the mating of individuals of the same species that are not closely related and hence are genetically dissimilar. It produces new combinations of characteristics and can result in hybrid vigour. *Contrast inbreeding* (↑).

P_1 Parental generation in breeding experiments on plants and animals. They are crossed to give offspring of the F_1 (↓) generation.

F_1 offspring or first filial generation produced by crossing plants and animals of the P_1 (↑) generation.

F_2 offspring or second filial generation produced by crossing plants and animals of the F_1 (↑) generation.

hybrid
production of hybrids by cross fertilization of two pure strains of garden pea

T = tall stems (dominant)
t = short stems (recessive)

	generation	genetic composition of parents	phenotype
1 crossing two pure strains			
	P_1	**TT** x **tt**	tall × short
gametes		**T** and **T** **t** and **t**	
	F_1	**Tt** **Tt**	tall (hybrids)
2 self pollination of F_1 generation			
	P_2	**Tt** x **Tt**	tall
gametes		**T** and **t** **T** and **t**	
	F_2	**TT** **Tt** **tT** **tt**	3 tall (TT, Tt, tT) 1 short (tt)

1 crossing two pure strains

| P₁ | | RR YY × rr yy | | round, yellow × wrinkled, green |

key:
R = round (dominant)
r = wrinkled (recessive)
Y = yellow (dominant)
y = green (recessive)

gametes — RY and RY × ry and ry

F₁ — Rr Yy × Rr Yy — all round, yellow

2 self-pollination of F₁ generation

P₂ — Rr Yy × Rr Yy — both round, yellow

Mendel's experiment on crossing plants with two pairs of contrasting characteristics

gametes — RY Ry rY ry × RY Ry rY ry
male gametes   female gametes

F₂	RY	Ry	rY	ry	
RY	RY RY round yellow	RY Ry round yellow	RY rY round yellow	RY ry round yellow	9 round, yellow
Ry	Ry RY round yellow	Ry Ry round green	Ry rY round yellow	Ry ry round green	3 round, green
rY	rY RY round yellow	rY Ry round yellow	rY rY wrinkled yellow	rY ry wrinkled yellow	3 wrinkled, yellow
ry	ry RY round yellow	ry Ry round green	ry rY wrinkled yellow	ry ry wrinkled green	1 wrinkled, green

Punnett's square

Mendel's laws

Mendel's laws (n.pl.) two laws formulated by Mendel to explain his experimental results on the transmission of inherited characteristics from parents to offspring. These are the Law of Segregation (↓) and the Law of Independent Assortment (↓).

Law of Segregation (n) the first of Mendel's laws (↑). It states that alleles segregate from one another during the formation of gametes (see diagram).

Law of Independent Assortment (n) the second of Mendel's laws (↑). It states that alleles of different genes are assorted independently of one another during the formation of gametes (see diagram).

cell division (n) process by which one cell splits to form two daughter cells. The nucleus divides first, by mitosis (↓), meiosis (↓) or, very rarely, by amitosis (↓). The cytoplasm then divides: in animal cells by constriction into two; in plants by the formation of a middle lamella.

binary fission (n) division into two. It is a common method of asexual reproduction in unicellular organisms, e.g. Amoeba. The nucleus of the parent cell undergoes mitosis (↓); this is followed by division of the cytoplasm and the formation of two daughter cells.

parent cell

nucleus of parent dividing

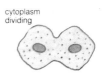
cytoplasm dividing

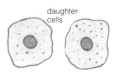
daughter cells

binary fission
binary fission in Amoeba

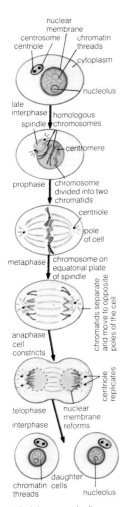

mitosis in an animal cell

mitosis

multiple fission (*n*) division into more than two, e.g. schizont of malaria divides into 16 merozoites.

multiply (*v*) to increase in numbers, e.g. cells multiply by cell division (↑).

amitosis (*n*) a type of nuclear division involving simple constriction of the nucleus into two halves. It is rare, found mostly in protozoa, e.g. macronucleus of ciliates.

mitosis (*n*) usual method by which plant and animal cell nucleus divides. The daughter nuclei are genetically identical with each other and with the parent nucleus. Mitosis occurs in four stages: prophase (↓), metaphase, anaphase and telophase (p.202). It is usually followed by division of the cytoplasm resulting in cell division.

meiosis, reduction division (*n*) type of division of the nucleus in animal and plant cells that results in halving of the chromosome number. It occurs in the formation of gametes in animals, production of pollen and the embryo sac of higher plants, formation of spores in plants that show alteration of generations and production of sexual spores in fungi. Meiosis has two functions: (1) results in the formation of four haploid daughter nuclei so when fusion of male and female haploid gametes occurs the zygote produced is diploid; (2) by segregation and crossing-over (p.207) it allows the offspring to inherit a mixture of maternal and paternal characteristics, i.e. the offspring are not identical with each other or with either parent. Meiosis involves two nuclear divisions in succession to yield four daughter nuclei: the **first meiotic division** takes place in four stages, prophase (↓) (sub-divided into leptotene (p.202), zygotene, pachytene, diplotene and diakinesis (p.203)), metaphase, anaphase and telophase (p.202). This is followed by the **second meiotic division** which resembles mitosis (↑) but does not usually involve a prophase because the chromosomes are already present as chromatids (p.203), i.e. it involves metaphase, anaphase and telophase.

prophase (*n*) (1) the first stage of mitosis (↑). The chromosomes become visible in the nucleus as long threads which then shorten, thicken, and partially divide into two identical chromatids (p.203). (2) the first stage of the first meiotic division (↑) which is sub-divided into leptotene (p.202), zygotene, pachytene, diplotene and diakinesis (p.203). Pairs of homologous chromosomes come together and divide into chromatids.

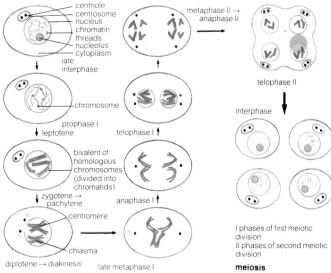

I phases of first meiotic
division
II phases of second meiotic
division

meiosis

metaphase (*n*) stage in mitosis or meiosis after
prophase (p.201). The nuclear membrane breaks down.
the nucleolus disappears and a spindle (↓) is formed.
The chromosomes line up on the equatorial plate.

anaphase (*n*) third stage of mitosis or meiosis (p.201). It
occurs after metaphase (↑). In mitosis and in the
second meiotic division the centromere (↓) divides, the
spindle (↓) elongates and the two chromatids (↓) move
to opposite poles. In the first meiotic division it is the two
homologous chromosomes that move to opposite poles.

telophase (*n*) last stage of mitosis or meiosis (p.201). It
occurs after anaphase (↑). In mitosis and in the second
meiotic division the chromatids become longer as they
uncoil and disperse, the spindle disappears, the nuclear
membrane reforms and the nucleolus reappears. In
telophase of the first meiotic division the chromatids do
not become dispersed and the second meiotic division
normally follows immediately. After telophase of both
mitosis and the second meiotic division the cytoplasm
divides and respectively 2 or 4 new cells are formed.

leptotene (*n*) first phase of prophase in the first meiotic
division (p.201). The chromosomes appear as thin
threads.

zygotene (*n*) second phase of prophase in the first meiotic division (p.201). Homologous chromosomes become associated to form a bivalent (↓); this process is known as pairing (↓).

pachytene (*n*) third stage of prophase in the first meiotic division (p.201). The bivalents (↓) coil and so become shorter and thicker. Each chromosome of the bivalent now seems to be divided longitudinally into two chromatids (↓) that remain joined by a centromere (↓).

diplotene (*n*) fourth phase of prophase in the first meiotic division (p.201). Crossing-over (p.207) occurs between two chromatids (↓), one derived from each of the original homologous pair of chromosomes.

diakinesis (*n*) fifth phase of prophase in the first meiotic division (p.201). Chromatids from each of the two homologous chromosomes begin to separate, other than at points where crossing-over has occurred. Diakinesis is followed by metaphase of the first meiotic division.

interphase (*n*) time when a nucleus is not undergoing mitosis or meiosis; there is intense biochemical activity during which DNA is replicated ready for the next division and energy is accumulated by the cell.

spindle (*n*) system of tubules formed in a cell during the metaphase (↑) of mitosis and meiosis. It is composed of a protein called tubulin. The chromatids (↓) are attached to the tubules by their centromeres (↓). The spindle changes shape at anaphase (*see diagrams*) and disappears at telophase.

equatorial plate (*n*) plane at the equator of spindle (↑). The chromatids (↓) are attached along it by their centromeres (↓) during the metaphase of cell division.

centromere, attachment constriction, spindle attachment (*n*) point of attachment to each other of the chromatids (↓) of a chromosome during mitosis or meiosis.

chromatid (*n*) one of two structures formed when a chromosome splits longitudinally during prophase (p.201) of mitosis or meiosis. They are joined at the centromere (↑). Chromatids separate at anaphase of mitosis and the second meiotic division; they are then called **daughter chromosomes**.

bivalent (*n*) two homologous chromosomes which pair during the zygotene phase of prophase in meiosis.

pairing (*n*) the coming together of two homologous chromosomes so that they lie side by side to form a bivalent (↑). It occurs during prophase of meiosis.

chromosome (*n*) a thread-like structure seen in the
nucleus of dividing cells. It contains DNA, the genetic
(↓) material of the cell. Chromosomes occur in pairs in
most nuclei; exceptions are gametes and cells of the
gametophyte generation of plants where there is only
one member of each pair per nucleus. The nuclei of one
species all contain the same set of chromosomes but
they may be in the haploid (↓), diploid (↓) or polyploid (↓)
state. Individual chromosomes in each set differ in
length, shape and position of the centromere. They
divide during mitosis and meiosis. The material that
makes up the chromosomes is chromatin (↓). The DNA
of chromosomes makes up the genes (↓) which occur in
a linear sequence along the chromosome. The order in
which the genes occur is called the **chromosome map**.
autosomes (*n*) the homologous pairs of chromosomes,
i.e. all those except the sex chromosomes (p.196).
homologous chromosomes (*n*) a pair of identical
chromosomes (↑) found in diploid (↓) organisms. They
pair at the prophase stage of meiosis.
chromatin (*n*) the material that makes up the
chromosomes (↑). It is composed of DNA and protein
and stains strongly with basic dyes.
haploid (*adj*) describes a nucleus, cell or organism that
has a single set of unpaired chromosomes (↑). The
number per nucleus is called the **haploid number** (in
man it is 23) and is half the diploid number (↓). Haploid
nuclei are normally formed as a result of meiosis. They
are found in gametes of plants and animals; gametophytes of plants; spores of Algae, Fungi and Bryophytes.
diploid (*adj*) describes a nucleus, cell or organism that
has two sets of chromosomes (↑). The number per nucleus is called the **diploid number** (in man it is 46) and
is twice the haploid number (↑). Diploid nuclei are found
in most animal cells except gametes; sporophytes of
plants; in some stages in the life of many Algae and
Fungi. They can undergo meiosis to form haploid nuclei.
triploid (*adj*) describes a nucleus, cell or organism with
three times the haploid (↑) number of chromosomes.
polyploid (*adj*) describes a nucleus, cell or organism with
three or more times the haploid (↑) number of chromosomes. It is fairly common in plants but rare in animals.
locus (*n*) the region on a chromosome (↑) occupied by a
particular gene. Alleles (↓) of a gene occupy corresponding positions on homologous chromosomes.

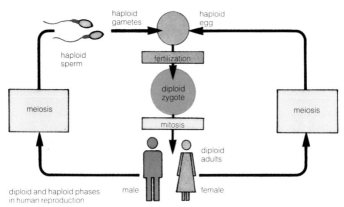

diploid and haploid phases in human reproduction

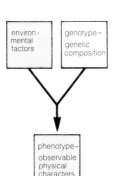

gene, hereditary factor (*n*) a unit of inherited material. It represents a small segment of a chromosome (↑) and is made up of DNA. The information it contains is a function of the order of its bases; *see genetic code* (p.209). This information is transferred to the ribosomes in the form of mRNA and is translated by them into a particular protein. Genes thus control the different characteristics of individuals. They are passed from one generation to the next by the sets of genes present in the gametes. Genes may be altered by mutation (p.207).

genetic (*adj*) concerned with genes (↑) and genetics (p.7).

alleles, allelomorphs (*n*) different forms of the same gene which occupy corresponding loci on homologous chromosomes (↑) but which express different, often contrasting characteristics. Usually there are only two alleles for each locus, (e.g those for tall and short-stemmed garden pea plants) but sometimes there may be three or more, i.e. **multiple alleles** e.g. there are three alleles giving rise to the ABO blood group system; these are labelled G^A, G^B, G. Note in these latter cases the usual diploid cell contains no more than two alleles. In each pair of alleles one is normally dominant (p.206), the other recessive (p.207).

genotype (*n*) the genetic composition of an organism. This is a function of the number and type of genes (↑) present in each cell. *Contrast phenotype* (↓).

phenotype (*n*) physical characteristics of an organism; affected both by the genotype (↑) and the environment.

homozygous (*adj*) describes a nucleus, cell or organism that has on homologous chromosomes (p.204) the same alleles at the locus or loci in question. The individual produces identical gametes with respect to the gene(s) in question. *Contrast heterozygous* (↓), *hemizygous* (↓).

generation	genetic composition			phenotype
1. crossing two pure strains				
P_1	**TT** ✕ **tt**			tall × short
gametes	**T** and **T**	**t** and **t**		
F_1	**Tt**	**Tt**		tall
2. self-pollination of F_1 generation				
P_2	**Tt** ✕ **Tt**			both tall
gametes	**T** and **t**	**T** and **t**		
F_2	**TT**	**Tt**	**tT**	**tt**	3 tall 1 short

☐ homozygous   ▨ heterozygous

heterozygous   homozygous

cross between two garden pea plants that are homozygous for tall and short stems

T = tall stems (dominant)
t = short stems (recessive)

heterozygous (*adj*) describes a nucleus, cell or organism that has on homologous chromosomes (p.204) different alleles at the locus or loci in question. The individual produces dissimilar gametes with respect to the gene(s) in question. *Contrast homozygous* (↑), *hemizygous* (↓).

hemizygous (*adj*) describes a nucleus, cell or organism in which a particular gene is not paired. This may be the result of the cell in question being haploid or the gene in question occupying a non-homologous region of a sex chromosome of a diploid organism. *Contrast homozygous* (↑), *heterozygous* (↑).

segregation (*n*) separation of alleles (p.205) on homologous chromosomes during meiosis so that each is in a different gamete; is the basis of Mendel's first law (p.200).

independent assortment (*n*) random assignment of genes to gametes at meiosis. It is the basis of Mendel's second law (p.200). It is not found in linked genes (↓).

dominant (*n*) in genetics, the allele which shows its effect in the phenotype not only in the homozygous state but also in the heterozygous state, e.g. in the garden pea there are two alleles that control the height of the stem. One, (T), gives a tall stem in the homozygous state, the other, (t), a short stem. In the heterozygous state (Tt) the plants are phenotypically tall showing that T is dominant. *Compare recessive* (↓).

mutation
mutations of a chromosome
containing genes A, B, C, D,
E, F

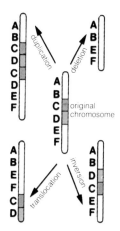

original
chromosome

chromatids of homologous
chromosomes which will
cross over

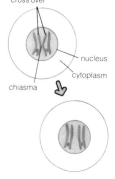

nucleus

cytoplasm

chiasma

crossing-over
crossing-over at the
diplotene stage of meiosis

recessive (*n*) in genetics, the allele which has no effect on the phenotype unless it is homozygous in the genotype; that is, it is present as a so-called **double recessive**. *Compare dominant* (↑).

mutation (*n*) usually a change in a segment of DNA (i.e. a gene) in a chromosome, often called a **gene mutation**. Mutations sometimes occur naturally but may be induced by chemicals, irradiation or neutron bombardment. Mutations can also arise from structural changes in chromosomes (i.e. **chromosome mutations**), e.g. deletion (↓), duplication (↓), translocation (↓), inversion (↓) and from changes in the number of chromosomes, e.g. polyploidy (p.204). Only if mutations are transmitted to gametes are they inherited by the offspring.

mutant (*n*) an organism containing one or more genes that have been altered by mutation (↑). Evolution occurs by natural selection of mutants.

deletion (*n*) type of mutation (↑) in which a segment of chromosome is lost during meiosis.

duplication (*n*) type of mutation (↑) in which a segment of chromosome is replicated twice during meiosis.

translocation (*n*) type of mutation (↑) in which a segment of chromosome is transferred either to another position on the same chromosome or to a different chromosome.

inversion (*n*) type of mutation (↑) in which a segment of chromosome is reversed so that genes within that part lie in reverse order.

crossing-over (*n*) the interchange of parts of chromatids at the diplotene stage of meiosis. The chromatids from paired homologous chromosomes break and rejoin in random combination. They form new combinations of genes on the chromosome and hence genetic variation in the gametes. The point at which crossing-over occurs is called a **chiasma** (*pl. chiasmata*). The formation of new assortments of genes as a result of crossing-over is referred to as **recombination**. The number of times this occurs is called **cross-over** frequency.

linkage (*n*) in genetics, refers to genes which occur on the same chromosome (i.e. **linked genes**). Because of this they do not show independent assortment (↑) at meiosis unless crossing-over (↑) has occurred.

sex-linked genes (*n*) genes which are carried on the X or Y chromosomes (i.e. the sex chromosomes p. 196).

DNA, deoxyribonucleic acid (*n*) the hereditary material of most living things (except some RNA containing viruses, e.g. influenza). The genetic information of the cell is stored in the order of its bases. DNA has a three-dimensional structure, consisting of two polynucleotide chains coiled together forming a **double helix**. Hydrogen bonds between the base pairs hold the structure together (*see diagram*). The nucleotides which make up each polynucleotide chain consist of a sugar, deoxyribose (↓); phosphoric acid; and one of four nitrogenous bases, adenine (↓), guanine (↓), cytosine (↓) or thymine (↓). It is the presence of deoxyribose and thymine which distinguishes DNA from RNA. DNA is replicated (p.210) prior to chromatid formation during meiosis or mitosis. It acts as the template both for the transcription (p.210) of RNA during protein synthesis and its own replication.

hydrogen bonds between bases

double helix of DNA

There are ten base pairs for every complete twist of helix
3–4 Å between base pairs
34 Å per twist
20 Å across

DNA
part of a DNA molecule

deoxyribose (*n*) a pentose sugar found in DNA (↑).
purine (*n*) one of a group of nitrogenous bases, e.g. adenine (↓) and guanine (↓), which are constituents of coenzymes (e.g. NAD), nucleotides (e.g. ATP) and nucleic acids (i.e. DNA and RNA).
pyrimidine (*n*) one of a group of nitrogenous bases, e.g. cytosine (↓), thymine (↓) and uracil (p.118) which are constituents of nucleotides and nucleic acids.
adenine (*n*) a purine (↑) base found in coenzymes (e.g. NAD, FAD), nucleotides (e.g. ATP) and nucleic acids.
guanine (*n*) a purine (↑) base found in some coenzymes and nucleic acids, i.e. DNA and RNA.
cytosine (*n*) a pyrimidine (↑) base found, for example, in nucleic acids, i.e. DNA and RNA.
thymine (*n*) a pyrimidine (↑) base found in DNA.

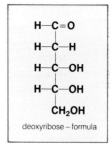

deoxyribose – formula

genetic code

U = uracil
A = adenine
G = guanine
C = cytosine

amino acid	triplet(s) which code for a particular amino acid					
alanine	GCU	GCC	GCA	GCG		
asparagine	AAU	AAC				
arginine	CGU	CGC	CGA	CGG	AGA	AGG
aspartic acid	GAU	GAC				
cysteine	UGU	UGC				
glutamic acid	GAA	GAG				
glutamine	CAA	CAG				
glycine	GGU	GGC	GGA	GGG		
histidine	CAU	CAC				
isoleucine	AUU	AUC	AUA			
leucine	UUA	UUG	CUU	CUC	CUA	CUG
lysine	AAA	AAG				
methionine	AUG					
phenylalanine	UUU	UUC				
proline	CCC	CCU	CCG	CCA		
serine	UCA	UCG	UCC	UCU	AGU	AGC
threonine	ACU	ACG	ACC	ACA		
tryptophan	UGG					
tyrosine	UAU	UAC				
valine	GUA	GUC	GUU	GUG		

genetic code (n) linear sequence of the bases in DNA (↑) of chromosomes. Each group (triplet) of three bases (which may be adenine, guanine, cytosine or thymine) codes for a particular amino acid. The code controls the inheritance from parents to offspring. It determines the order of nucleotides in a molecule of messenger RNA which in turn decides the order in which amino acids are linked in the synthesis of proteins. All organisms apparently use the same genetic code.

base pairing (n) the linking, by hydrogen bonds, of purine and pyrimidine bases (↑) in opposite strands of the double helix of DNA. Adenine is always linked with thymine; guanine with cytosine.

base ratio (n) ratio in DNA (↑) of the number of adenine (A) and thymine (T) bases to the number of cytosine (C) and guanine (G). It often varies between species, but in any one A=T, G=C.

base composition (n) in DNA (↑) the amount of guanine plus cytosine relative to the total amount of nucleotide, expressed as a percentage. In man the value is about 40%.

replication (*n*) process in which DNA acts as a template for the synthesis of another DNA molecule. In this way the genetic information stored in DNA is passed on accurately to the new DNA.

transcription (*n*) process in which DNA acts as a template for the manufacture of molecules of messenger RNA. In this way genetic information stored in DNA is used to determine the amino acid sequence in a protein.

translation (*n*) process in which the genetic code on messenger RNA determines the sequence in which the amino acids are linked in protein synthesis. It occurs on ribosomes.

life cycle (*n*) succession of changes through which an organism passes in its development from a fertilized egg to a mature individual which can produce the next generation. In simple life cycles an organism produces an offspring similar to itself (e.g. in man). In more complex life cycles intermediate stages may exist (e.g. butterfly). The most complex life cycles, found for example in parasites like the malarial parasite, involve a large number of stages in more than one host (↓).

parasite (*n*) an animal or plant that lives on or in, and obtains food and sometimes protection from, another living organism, the host (↓) to which it usually does harm. They may be ectoparasites (↓) or endoparasites (↓). **Obligate parasites** can live only parasitically; **facultative parasites** can also live as saprophytes. Some parasites, e.g. the trypanosome and schistosome, have complex life cycles (↑) involving more than one host (↓). **Partial parasites**, e.g. mistletoe, can carry out photosynthesis but obtain some nutrients from the host; **complete parasites,** e.g. dodder, obtain all nutrients from the host.

ectoparasite (*n*) parasite (↑) that lives on the external surface of the host (↓), e.g. flea, louse.

endoparasite (*n*) parasite (↑) that lives within the host body, e.g. malarial parasite, tapeworm, schistosome.

host (*n*) (1) organism that provides a parasite (↑) with food and protection. A **primary** or **definitive host** is that in which the parasite becomes sexually mature. A **secondary** or **intermediate host** is that which supports the asexual stages of the parasite. (2) any organism that provides physical support for another. (3) an animal onto which a graft is transplanted.

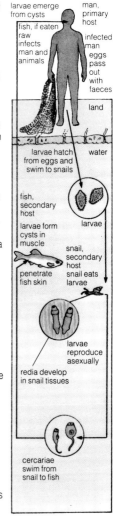

larvae emerge from cysts

man, primary host

fish, if eaten raw infects man and animals

infected man

eggs pass out with faeces

land

larvae hatch from eggs and swim to snails

water

fish, secondary host

larvae form cysts in muscle

larvae

penetrate fish skin

snail, secondary host

snail eats larvae

larvae reproduce asexually

redia develop in snail tissues

cercariae swim from snail to fish

host
hosts of the liver fluke

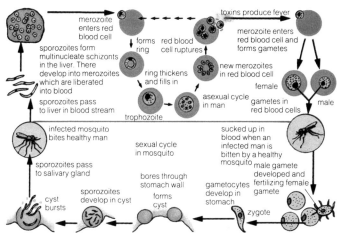

merozoite enters red blood cell

sporozoites form multinucleate schizonts in the liver. There develop into merozoites which are liberated into blood

sporozoites pass to liver in blood stream

toxins produce fever

merozoite enters red blood cell and forms gametes

forms ring red blood cell ruptures

new merozoites in red blood cell

ring thickens and fills in

female

asexual cycle in man

gametes in red blood cells

male

trophozoite

infected mosquito bites healthy man

sexual cycle in mosquito

sucked up in blood when an infected man is bitten by a healthy mosquito

sporozoites pass to salivary gland

bores through stomach wall

male gamete developed and fertilizing female gamete

gametocytes develop in stomach

cyst bursts

sporozoites develop in cyst

forms cyst

zygote

life cycle of the malarial parasite

vector, disease vector (*n*) any agent which transmits the causative agent (↓) of a disease (p.214). Physical vectors are air, water, food and contact. Animal vectors include mosquito (spreads malaria and yellow fever), blackfly (spreads onchocerciasis), dog (spreads rabies). Examples of some common diseases and their vectors are given on the back endpaper. Some diseases have more than one vector, e.g. infective hepatitis. The precise vectors of others are not known, e.g. poliomyelitis, leprosy.

causative agent (*n*) the organism, usually a micro-organism, that causes disease (p.214). It may be a virus (p.26), rikettsia (p.26), bacterium (p.27), spirochaete (p.27), protozoan (p.21), fungus (p.19, 32), worm (p.31). Examples of some common diseases and their causative agents are given on the back endpaper.

malaria (*n*) disease of man; the causative agent is a protozoan called ***Plasmodium*** or the **malarial parasite**. It is transmitted to man by the bite of an infected ***Anopheles* mosquito** (the vector). *Plasmodium* causes damage to the red blood cells of man and releases toxins which cause fever periodically every 48 or 72 hours depending on the species of *Plasmodium*. The parasite has a complicated life cycle showing two phases, an asexual phase in man and a sexual phase in the mosquito.

trypanosomiasis (**sleeping sickness** in man; **nagana** in cattle) (*n*) a disease caused by a parasite protozoan *Trypanosoma*. Some species affect man, others affect cattle. The parasite has a complicated life cycle showing two phases, one in man/cattle, the other in the tsetse fly. The disease is transmitted by the bite of an infected tsetse fly (the vector).

cyst (*n*) (1) outer structure encasing the resting stage of an organism; seen, for example, in *Entamoeba* and liver fluke (↓). (2) a bladder or sac-like structure in body tissues that encloses a liquid, usually formed as a result of inflammation.

encystation (*n*) process of cyst (↑) formation.

schistosomiasis, Bilharzia, Bilharziasis (*n*) a disease, normally of man, caused by an infestation of parasitic flatworms or blood flukes (↓) called *Bilharzia* or *Schistosoma*. The flukes have a complicated life cycle (*see diagram*). Adult worms live in man but in order to complete the cycle the flukes must undergo some stages of development in a particular species of snail (the secondary host). The larvae emerging from the snail, the **Cercariae,** penetrate the skin of man.

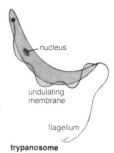

nucleus

undulating
membrane

flagellum

trypanosome

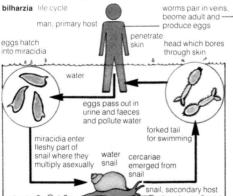

bilharzia life cycle

man, primary host

eggs hatch
into miracidia

water

worms pair in veins,
beome adult and
produce eggs

penetrate
skin

head which bores
through skin

eggs pass out in
urine and faeces
and pollute water

forked tail
for swimming

miracidia enter
fleshy part of
snail where they
multiply asexually

water
snail

cercariae
emerged from
snail

snail, secondary host

female

male

fluke (*n*) parasitic flatworm, e.g. *Fasciola* (liver fluke), *Schistosoma* (blood fluke) (↑). The schistosome has a complicated life cycle with more than one host involving larval stages in the snail and adult stages in man. Infection from snails is by skin penetration of cercariae released from infected snails.

venereal diseases, VD (*n*) a group of highly infectious diseases of man normally spread by sexual contact. The two most common are **syphilis,** caused by a spirochaete and **gonorrhoea,** caused by a gonococcus. Both can be treated with antibiotics, e.g. penicillin.

drug (*n*) a chemical substance used for medical or veterinary purposes. Drugs are either used to control diseases caused by infective agents (e.g. **chloramphenicol** and **chloromycetin** used to treat typhoid; **quinine** and **chloroquine** used in treatment of malaria) or to regulate body functions (e.g. epinephrine, a bronchial dilator used to treat asthma; **heparin,** an anticoagulant used to treat coronary heart disease; aspirin used to relieve pain). Some drugs, e.g. narcotics (↓), can be harmful if used incorrectly.

antibiotic (*n*) a drug (↑) produced by micro-organisms, usually bacteria or fungi, which is secreted into the growth media. Antibiotics are used in medicine to treat many microbial infections, e.g. **penicillin** is a group of substances extracted from moulds, particularly *Penicillium notatum;* **tetracyclines,** extracted from cultures of *Streptomyces,* are active against a wide range of bacteria and rickettsiae.

narcotic (*n*) any substance, e.g. a drug (↑), alcohol, that affects the central nervous system causing dizziness, sleep, loss of memory and eventually unconsciousness. Narcotics are used in medicine to relieve pain. They are dangerous in excess because they cause hallucination, convulsions and death. Some, e.g. **morphine** and **opium**, are particularly dangerous as they are addictive.

synergism (*n*) interaction in one particular system of two or more substances, e.g. drugs (↑), so that they act together to produce a greater effect than the sum of the individual effects. *Contrast antagonism* (↓).

antagonism (*n*) (1) interaction in one particular system of two or more substances, e.g. drugs, hormones, so they act against one another to produce a smaller effect than the sum of the individual effects. *Contrast synergism* (↑). (2) interaction of two closely associated structures or organisms so the effect of one is inhibited, e.g. two muscles acting in opposition show antagonism.

antiviral, antibacterial, antiprotozoal, antiparasitic, antifungal, anthelminthic drugs (*n*) drugs acting respectively against viruses, bacteria, protozoa, parasites, fungi and helminths.

disease (*n*) disorder of a tissue, organ or organism. Diseases may be caused by (1) infectious agents (infectious diseases (↓)), e.g. viruses, bacteria, protozoa, parasites, worms, fungi or by (2) various internal physiological defects where no external agent is involved, e.g. angina, diabetes. Infectious diseases are spread by a variety of vectors; examples of some together with the causative agents of the disease are given on the back endpaper. All diseases are identified by their signs and symptoms (p.216).

infectious disease (*n*) disease (↑) due to the presence of a pathogen (e.g. virus, bacterium, protozoan, parasite, worm, fungus). They are often spread from person to person by vectors (e.g. air, water, animals).

contagious disease (*n*) infectious disease (↑), particularly of the skin, that is spread by personal contact, e.g. ring worm.

congenital disease (*n*) disease (↑) or deformity an animal already has when it is born, e.g. haemophilia.

acute disease (*n*) disease (↑) which runs its course rapidly and is often fatal, e.g. plague. *Contrast chronic disease* (↓).

chronic disease (*n*) disease (↑) which continues for a long time. The condition either remains static or only increases in severity very slowly, e.g. Chagas' disease. *Contrast acute disease* (↑).

lethal or **fatal disease** (*n*) disease (↑) from which recovery is unlikely without medical intervention.

endemic (*adj*) describes (1) a disease (↑) which is regularly found in a group of people or in a particular district. (2) pests or disease-causing parasites (e.g. mosquitos) that are regularly found in a particular area.

epidemic (*adj*) describes a disease (↑) which suddenly attacks, at one time, large numbers of people in a particular area.

pandemic (*adj*) describes an epidemic (↑) occurring simultaneously over a wide area, e.g. a continent.

sporadic (*adj*) describes a disease (↑) that occurs from time to time in widely separated places.

deficiency disease (*n*) disease (↑) caused by lack of one or more nutrients in the diet, e.g. kwashiorkor (↓), marasmus (↓), PCM (↓).

kwashiorkor (*n*) deficiency disease (↑) found in young children; caused by insufficient protein or the incorrect type of protein in the diet.

marasmus (*n*) deficiency disease (↑) in man caused by insufficient carbohydrate in the diet. This may be the result of intestinal infections (↓) or worm infestations (↓).

protein-calorie-malnutrition, PCM (*n*) deficiency disease (↑) in young children. It shows some symptoms of both kwashiorkor (↑) and marasmus (↑) and is caused by insufficient carbohydrate and protein in the diet.

pathogen

poliomyelitis virus

pathogen (*n*) an organism that causes a disease (↑), for example parasites like viruses, bacteria, protozoa and worms. Pathogens may be pathogenic to one organism but not to another type of organism. **pathogenic** (*adj*).

virulence (*n*) the degree to which a parasite is pathogenic (↑), i.e. causes disease.

virulent (*adj*) describes a pathogen (↑) that causes severe disease.

smallpox virus

infection (*n*) (1) process by which pathogens (↑) enter a tissue, organ or organism; (2) an infectious disease (↑).

infect (*v*) to cause an infection (↑).

syphilis
spirochaetes

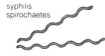

infestation (*n*) infection (↑) caused by large parasites, particularly fleas, lice, ticks, worms; e.g. schistosomiasis is an infestation of a flatworm, *Schistosoma mansoni*.

infest (*v*) to cause an infestation (↑).

reservoir of infection (*n*) a residual source of a pathogen (↑) after the disease it causes has been cured in a human population; e.g. animals and man can both suffer from certain diseases. Where this is controlled in man, but not in animals the animals act as a reservoir of infection. The appropriate animal vector can then transmit the disease back to man.

cholera
bacteria

carrier (*n*) an infected person who transmits a disease but does not appear to suffer from it, i.e. he shows no signs or symptoms; e.g. a typhoid carrier appears healthy but has bacteria causing typhoid in his gut and can infect food he handles and so spread the disease.

quarantine (*n*) the keeping in isolation of men or animals that may have been in contact with people suffering from an infectious disease (e.g. smallpox, cholera, plague, yellow fever, rabies), until it is known whether they have become infected. This is to prevent the spread of infection. The quarantine period must exceed the incubation period (p.216) of the disease.

isolation (*n*) the separation of a person or animal, known to be suffering from an infectious disease (↑), from the rest of the community. This is to prevent the disease spreading. **isolate** (*v*).

incubation period (*n*) time between the entrance of a
pathogen (p.215) and the first clinical symptoms (↓) of
disease in the host.

infective period (*n*) period of time that a pathogen can
be transmitted from one infected person to another.

signs of disease (*n*) visible, outward indication of the
presence of disease. For example, in man these may
include pallor (↓), rash (↓), jaundice (p.131),
inflammation (↓), oedema (↓). Signs and symptoms (↓)
are considered together in the diagnosis (↓) of disease.

symptoms of disease (*n*) effects felt by a person as a
result of disease or injury, e.g. fever (↓), pain, headache,
insomnia, fatigue (↓). Signs (↑) and symptoms are
considered together in the diagnosis (↓) of disease.

diagnosis (*n*) the identification of a disease by the
nature of its signs and symptoms (↑). In man, this is
normally performed by a doctor; in animals, by a
veterinary surgeon (vet). **diagnose** (*v*).

pallor (*n*) paleness of the skin. It can be a sign (↑) of
shock or disease.

rash (*n*) a large number of spots on the skin. It may be a
sign (↑) of disease or the result of an allergy (↓).

inflammation (*n*) a localised reaction of the body in
response to injury. The tissues in the injured area swell
and redden because of dilation of blood vessels. These
effects are produced by release of histamine (↓) from
damaged cells. Inflammation can also result from
allergic reactions (↓).

oedema (*n*) (1) in animals, a swelling due to an
accumulation of fluid in tissue spaces, especially under
the skin. It is a sign (↑) of disease. The condition is also
known as **dropsy**. (2) in plants, a mass of unhealthy,
swollen parenchyma.

contusion (*n*) a bruise, i.e. a discoloration of a person's
skin caused by a blow; it is often accompanied by swelling. The skin surface is unbroken but blood vessels
under it are ruptured making it appear blue/black.

trauma (*n*) (1) a wound or injury. (2) the state caused by
a physical or emotional shock.

fever (*n*) (1) a symptom (↑) of disease in animals. It
involves an increase in body temperature from the
normal and is usually accompanied by a more rapid
pulse rate. (2) a particular disease, usually infectious,
with feverish symptoms, e.g. scarlet fever, glandular
fever.

rash

human skin with rash
caused by chicken pox

large numbers
of spots

contusion
contusion of upper
limb of man

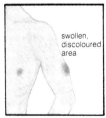

swollen,
discoloured
area

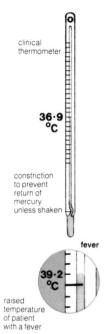

clinical
thermometer

36·9
°C

constriction
to prevent
return of
mercury
unless shaken

fever

39·2
°C

raised
temperature
of patient
with a fever

measurement of temperature
of the human body

histamine
formula of histamine

shiver (*v*) to shake with involuntary, muscular movements which may be due to low environmental temperature or the presence of a fever (↑).

vomit (*v*) to forcibly expel the contents of the stomach through the mouth. This may have a variety of causes. e.g. the presence of toxins or pathogens in the alimentary canal, or emotional stress.

nausea (*n*) an unpleasant desire to vomit (↑). It may have a variety of causes, e.g. disease, stomach disorder, abnormal motion of the body as in travel sickness.

fatigue (*n*) excess tiredness, normally relieved by a period of rest.

paralysis (*n*) loss of sensations and the power of movement in a part of the body. It is caused by lack of impulses along nerves or failure of neurotransmitter. It may be due to disease, e.g. poliomyelitis or accidental damage to the central nervous system, e.g. fracture of the spine.

convulsion (*n*) a violent, involuntary contraction (or spasm) of the voluntary muscles of the body. It is caused by damage to the brain.

epilepsy (*n*) a disease of the brain (possibly inherited). It involves recurrent, sudden attacks of unconsciousness, often accompanied by convulsions (↑).

diarrhoea (*n*) frequent passage of abnormal, usually liquid, faeces from the bowel; often the result of disease, e.g. bacillary dysentery. *Contrast constipation* (↓).

constipation (*n*) irregular and insufficient passage from the bowel of faeces which are usually hard and dry. *Contrast diarrhoea* (↑).

allergy (*n*) result of abnormal reactions of body tissues to substances like pollen, animal fur, dust, certain foods, drugs, e.g. penicillins, which are normally harmless to most other individuals. Symptoms can be relieved by antihistamine drugs (↓).

histamine (*n*) an amine present in body tissues which is released into the blood from the mast cells (p.223) when tissues are injured. It causes local inflammation. Histamine release can also be caused by allergic reactions (↑).

antihistamine drug (*n*) one of a group of drugs that antagonises the action of histamine (↑) in allergic conditions (↑).

pus (*n*) thick, yellowish fluid consisting of serum containing dead white blood cells and bacteria.

tumour, tumor (*n*) lump in the body, without inflammation. It is caused by an abnormal growth of cells. It may be due to the presence of an infectious organism or it may occur spontaneously. In the latter case it can be benign (↓) or malignant (↓). Tumours may cause pain and malfunction by pressing on other tissues and organisms.

neoplasm (*n*) an abnormal new growth, e.g. tumour (↑).

cancer (*n*) any malignant (↓) new growth or malignant tumour (↑), e.g. carcinoma (↓), sarcoma (↓). Cancers invade neighbouring tissues and can spread in the lymphatics and blood vessels to other areas.

carcinoma (*n*) a cancer (↑) which is a malignant growth of epithelial tissue.

sarcoma (*n*) a cancer (↑) which is a malignant growth of connective tissue.

carcinogen (*n*) any substance or agent that encourages the growth of a cancer (↑), e.g. certain chemicals.

malignant (*adj*) describes (1) an abnormal growth which can often spread to other areas and may eventually cause death, e.g. a cancer (↑); (2) a disease which becomes worse and may be fatal. *Compare benign* (↓).

benign (*adj*) describes (1) an abnormal growth, e.g. a tumour, which does not spread, i.e. it is not malignant (↑); (2) a mild, non-fatal disease. *Compare malignant* (↑).

transplantation (*n*) (1) transfer of an organ or tissue from an animal, either to another position on the same animal, or to another animal. The part transplanted is called a graft (p.79), the individual supplying it is the donor (↓) and the individual receiving it is the recipient (↓), e.g. a kidney transplant (2) transfer of a plant from one place to another, e.g. seedlings are transplanted from a nursery to open ground.

donor (*n*) an animal that provides a tissue or organ for transplantation (↑) or supplies blood for transfusion (↓).

recipient (*n*) an animal that either receives a tissue or organ from another individual by transplantation (↑) or who receives blood during transfusion (↓).

transfusion (*n*) transfer into a vein of a fluid such as plasma, serum, blood, saline, glucose.

blood transfusion (*n*) transfusion (↑) of blood from a donor (↑) into the body of a recipient (↑). Blood must be of the correct blood group (see p.141), i.e. it must be compatible (↓). Blood is transfused to replace diseased blood or that lost as a result of accident or surgery (↓).

transplantation

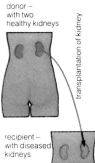

donor –
with two
healthy kidneys

transplantation of kidney

recipient –
with diseased
kidneys

donor
recipient
donors and recipients
in blood transfusions

recipient blood group →

	O	A	B	AB
O	✓	✓	✓	✓
A	✕	✓	✕	✓
B	✕	✕	✓	✓
AB	✕	✕	✕	✓

↑
donor blood group

▶ universal recipient
▷ universal donor
✓ compatible transfusion
✕ incompatible transfusion

sphygmomanometer
operation of a
sphygmomanometer

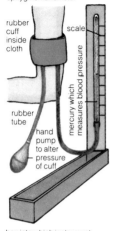

rubber
cuff
inside
cloth

scale

mercury which
measures blood pressure

rubber
tube

hand
pump
to alter
pressure
of cuff

box into which instrument
is folded after use

ear
pieces
which
fit in
doctor's
ears

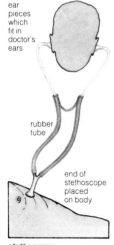

rubber
tube

end of
stethoscope
placed
on body

stethoscope

universal donor (*n*) person of blood group O, whose blood can be transfused into individuals of all blood groups, i.e. O, A, B, AB.

universal recipient (*n*) person of blood group AB, who can receive at a transfusion (↑) blood from donors of all blood groups, i.e. O, A, B, AB. (see p.141).

compatibility (*n*) ability to exist together, e.g. compatibility between donor and recipient is necessary for transplantation (↑) and blood transfusion (↑) so the new tissue is not rejected. *Contrast incompatibility* (↓).

incompatibility (*n*) inability to exist together, e.g. incompatibility between donor and recipient in transplantation (↑) and blood transfusion (↑) leads to rejection of the new tissue. In transplant operations this is minimised by a combination of **tissue typing** (i.e. tissue matching) and the use of drugs. *Contrast compatibility* (↑).

intravenous (*adj*) describes something introduced into a vein, e.g. an injection, a tube for diagnostic purposes.

sphygmomanometer, sphygmometer (*n*) an instrument, particularly used in human medicine, to measure arterial blood pressure.

stethoscope (*n*) an instrument, particularly used in human medicine, to listen to sounds produced within the body; e.g. sounds produced by the heart and lungs can be heard by placing the stethoscope against the body – these help in clinical diagnosis (↓) of disease.

clinical diagnosis (*n*) the identification of a disease by observation of its signs and symptoms (p.216).

clinical analysis (*n*) analysis of a substance or tissue (e.g. urine, blood, cerebrospinal fluid) from an animal or human. It helps in diagnosis (p.216) of disease.

surgery (*n*) (1) a branch of medical science, practised by a surgeon; it concerns the treatment of injury or disease by manual operations; (2) the consulting room of a doctor or dentist.

pharmacy (*n*) (1) a branch of medical science concerned with the knowledge of drugs and their use in medicines; (2) a place where drugs are prepared for a patient, also called a **dispensary**.

pharmacology (*n*) study of the distribution, effects, metabolism and excretion of drugs in animals, including man.

chemotherapy (*n*) treatment of a disease by use of a chemical compound (i.e. a drug p. 213) that will affect the particular pathogen involved, either by causing its destruction or by preventing its multiplication.

prophylaxis (*n*) any means to prevent the spread of disease. There are four main types: (1) improvements in the standard of housing and hygiene, for example, use of disinfectants, provision of a piped water supply and sewage collection system; (2) **chemoprophylaxis** – use of drugs, e.g. daraprim for malaria; (3) **immunoprophylaxis** – use of vaccines, e.g. for smallpox; (4) where appropriate destruction of, or avoidance of contact with, vectors, e.g. malaria, use of insecticides and mosquito nets; Bilharzia, destruction of snails and avoidance of physical contact with water which is likely to be contaminated.

toxin (*n*) a poison produced by a living organism that is harmful not to itself but to other organisms, e.g. diphtheria toxin. Toxins are usually proteins. **toxic** (*adj*).

toxicity (*n*) the extent to which a substance is toxic (↑).

venom (*n*) poison produced by an animal; it is usually injected into another animal by a bite or sting, e.g. snake venom is transmitted by a snake bite.

toxoid (*n*) a toxin (↑) that has been treated to remove its toxic properties but which is still able to stimulate the formation of antibodies.

disinfect (*v*) to destroy pathogens, particularly bacteria, found outside a human or animal body, i.e. in its immediate surroundings. This helps prevent infection. It involves the use of chemicals called **disinfectants** which are stronger than antiseptics (↓) and not suitable for use on the skin or for internal consumption.

disinfest (*v*) to remove an infestation (p.215) of animals, e.g. fleas and lice (1) from the body of another animal or human; (2) from a building. This is often carried out by fumigation (↓).

fumigation (*n*) the exposure to fumes, e.g those of the gas sulphur dioxide, in order to disinfest (↑) and kill parasites and pests. **fumigate** (*v*).

antisepsis (*n*) destruction or inhibition of the growth of bacteria, particularly in infected wounds. This is usually by the use of chemicals called **antiseptics**, e.g. iodine solution. Antiseptics are milder than disinfectants (↑).

asepsis (*n*) exclusion of micro-organisms, e.g. by sterilization of all objects used in surgery. **aseptic** (*adj*).

pasteurization (*n*) sterilization of liquids by raising the temperature high enough to kill bacteria but not high enough to damage the liquid; e.g. milk is pasteurized to kill bacteria causing tuberculosis and brucellosis.

prophylaxis

spraying of insecticide

immunisation

healthy person

disinfectant

suppressive drugs

boiling water

antiseptic

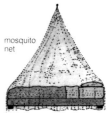

mosquito net

general prophylactic measures

immunity (*n*) systems, especially in animals, concerned with the recognition and destruction of foreign materials (antigens p.222) which enter the body. These antigens may be life-threatening infectious micro-organisms (e.g. viruses, bacteria, protozoa, fungi, helminths) or they may be life-saving transplants such as kidneys and hearts. The results of having an immune system may be desirable, e.g. natural resistance to infection, recovery from infection and the gaining of resistance to reinfection, or they may be undesirable, e.g. rejection of transplants and transfused blood, autoimmunity (p.224), hypersensitivity (p.224). Two main types of immunity occur; natural immunity (↓) and acquired immunity (p.222).

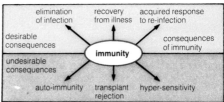

natural immunity, non-specific immunity (*n*) one of two types of immunity (↑). Natural immunity is inborn and not changed by the antigens to which the animal has been exposed; *compare adaptive immunity* (p.222). It involves humoral factors (e.g. interferon, lysozyme (p.222), complement (p.223)) and/or cells (mast cells, polymorphs, macrophages). It shows no specificity to any particular antigen. *Compare specific immunity* (p.222).

natural immunity
types of immunity

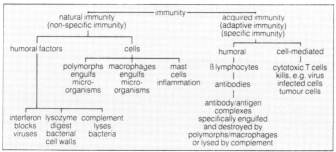

**acquired immunity, adaptive immunity, specific
immunity** (*n*) one of two types of immunity (p.221).
Acquired immunity involves B- and T-lymphocytes (↓)
which react specifically and selectively to the presence
of antigens by forming respectively antibodies in
humoral immunity (↓) or cytotoxic T-cells in cell-
mediated immunity (↓). These sometimes function by
interaction with parts of the natural immune system
(p.221). The response of the body in acquired
immunity, unlike natural immunity, is altered by the
antigens to which the animal has been exposed before;
see immunological memory (p.224). Acquired immunity
responds both to the antigens of natural infections
(naturally acquired immunity (↓)) and to those present in
vaccines (artificially acquired immunity (↓)).

naturally acquired immunity (*n*) acquired immunity (↑)
resulting from a natural infection by a disease-causing
micro-organism. *Contrast artificially acquired immunity* (↓).

artificially acquired immunity (*n*) acquired immunity (↑)
which results from the injection into an animal of antigens from disease-producing micro-organisms during
the process of immunisation or vaccination (p.224).

passive immunity (*n*) type of acquired immunity (↑)
which results from the transfer normally of antibodies
but occasionally of cytotoxic T-cells (↓) from an immune
to a non-immune animal. Occurs naturally between
mother and foetus via the placenta or mother and baby
via milk. Before antibiotics were in common use it was
also carried out artificially to protect humans suffering
from some bacterial infections.

immune system (*n*) system in an animal responsible for
immunity, especially of the acquired type. It can be
divided into two types: **humoral** which involves
production of antibodies and **cell-mediated** which
involves production of cytotoxic T-cells.

antigen (*n*) material to which the immune system (↑)
responds (see p.141). Antigens may be life-threatening
infectious,micro-organisms, materials in the
environment (*see hypersensitivity* p.224) or life-saving
transplants such as kidneys and hearts.

interferon (*n*) part of the non-specific immune system
(p.221); see p.26.

lysozyme (*n*) part of the non-specific immune system
(p.221). It is secreted by macrophages (p.144).
Present in tears, mucosa of respiratory tract.

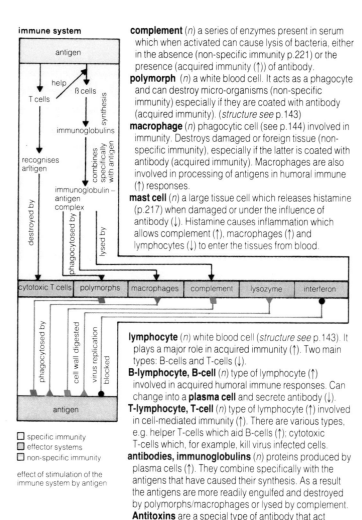

immune system

specific immunity
effector systems
non-specific immunity

effect of stimulation of the
immune system by antigen

complement (*n*) a series of enzymes present in serum
which when activated can cause lysis of bacteria, either
in the absence (non-specific immunity p.221) or the
presence (acquired immunity (↑)) of antibody.

polymorph (*n*) a white blood cell. It acts as a phagocyte
and can destroy micro-organisms (non-specific
immunity) especially if they are coated with antibody
(acquired immunity). (*structure see* p.143)

macrophage (*n*) phagocytic cell (see p.144) involved in
immunity. Destroys damaged or foreign tissue (non-
specific immunity), especially if the latter is coated with
antibody (acquired immunity). Macrophages are also
involved in processing of antigens in humoral immune
(↑) responses.

mast cell (*n*) a large tissue cell which releases histamine
(p.217) when damaged or under the influence of
antibody (↓). Histamine causes inflammation which
allows complement (↑), macrophages (↑) and
lymphocytes (↓) to enter the tissues from blood.

lymphocyte (*n*) white blood cell (*structure see* p.143). It
plays a major role in acquired immunity (↑). Two main
types: B-cells and T-cells (↓).

B-lymphocyte, B-cell (*n*) type of lymphocyte (↑)
involved in acquired humoral immune responses. Can
change into a **plasma cell** and secrete antibody (↓).

T-lymphocyte, T-cell (*n*) type of lymphocyte (↑) involved
in cell-mediated immunity (↑). There are various types,
e.g. helper T-cells which aid B-cells (↑); cytotoxic
T-cells which, for example, kill virus infected cells.

antibodies, immunoglobulins (*n*) proteins produced by
plasma cells (↑). They combine specifically with the
antigens that have caused their synthesis. As a result
the antigens are more readily engulfed and destroyed
by polymorphs/macrophages or lysed by complement.
Antitoxins are a special type of antibody that act
against bacterial toxins. Immunoglobulins (abbreviation
Ig) occur in five classes: Ig G, Ig M, Ig A, Ig D, Ig E and
have two or five sites with which they can combine with
antigen.

immune response (*n*) sequence of events that occurs when the immune system of an animal is activated by the presence of antigen. There are two types: **primary immune responses** are those to antigens to which the animal has not been exposed before; **secondary immune responses,** which occur more quickly and produce higher levels of antibody/cytotoxic T-cells, are to those antigens the animal has been exposed before.

immunological memory (*n*) a primary immune response (↑) produces not only antibodies or cytotoxic T-cells but also a type of lymphocyte called a 'memory cell', which ensures that a secondary immune response (↑) occurs if the antigen appears again.

vaccination, immunisation (*n*) injection of antigen in the form of killed or attenuated (↓) micro-organisms into an animal/man so that if a natural infection occurs later it will be cleared before clinical symptoms of the disease are evident. Although it raises serum antibody levels, vaccination works mainly through immunological memory (↑) by making sure that subsequent infections provoke secondary immune responses (↑).

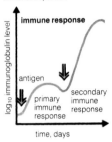

primary and secondary immune responses

vaccination

vaccines, their nature and length of immunity conferred

disease	nature of vaccine	period of immunity
small pox	living attenuated vaccinia virus	3 years
poliomyelitis	living attenuated virus (by mouth)	6 months
yellow fever	living attenuated virus	10 years
rabies	living attenuated virus	1 year
influenza	killed virus (by formalin)	6 months
typhus	killed rickettsiae	6 months

disease	nature of vaccine	period of immunity
plague	living attenuated bacillus	6 months
tuberculosis	living attenuated bacillus	4 years
typhoid	killed bacillus	1 year
cholera	killed bacillus	6 years
diphtheria	toxoid	3 years
tetanus	toxoid	6 years

attenuation (*n*) treatment of a culture of micro-organisms so its antigenicity is retained but its infectivity is much reduced or lost. It is then used as a vaccine (↑).

tolerance (*n*) failure of the immune system of an animal to react to its own body materials.

autoimmunity (*n*) reaction of the immune system of an animal with its own body materials due to a breakdown of tolerance(↑). This can lead in man to autoimmune diseases, e.g. pernicious anaemia.

rejection (*n*) in immunology, destruction of a tissue or organ transplant, usually due to an acquired cell-mediated immune response.

hypersensitivity (*n*) damage to tissues around an antigen due to over-reaction of components of the immune system (p.222), e.g. hayfever.

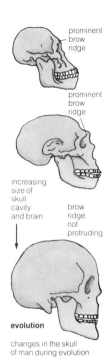

prominent
brow
ridge

prominent
brow
ridge

increasing
size of
skull
cavity
and brain

brow
ridge
not
protruding

evolution

changes in the skull
of man during evolution

Charles Darwin
Darwinism

evolution (n) the process by which more complex forms of life have arisen from simpler forms over millions of years. (*See geological time scale*, p.226 *and natural selection* (↓).)

phylogeny (n) history of development during evolution(↑).

natural selection (n) the method by which evolution (↑) is believed to have occurred. It arises from the facts that: (1) there is overproduction of offspring; (2) individuals within a population vary; (3) resources in terms of food, territory, etc., are limited; (4) only the fittest survive and produce the next generation. If the characteristics that aid survival are inherited the composition of the population will gradually change, i.e. successful individuals and their offspring are selected; non-successful ones are gradually eliminated. This evolutionary method was first proposed by Darwin.

Darwinism (n) the theory of evolution (↑) based on natural selection (↑), as proposed by Charles Darwin in 1859. It was formulated after his voyage in the *HMS Beagle*. (*See natural selection* (↑), *sexual selection* (↓).)

sexual selection (n) type of natural selection (↑) based on the preference of females for mates with particular sexual characteristics. It was suggested by Darwin in an attempt to explain exaggerated sexual character- istics seen in many animals, e.g. mane in lion.

adaptation (n) a change in a characteristic of an organ- ism that improves its chances of survival and producing offspring in a particular environment. Over many generations it results in evolution (↑) of a new species.

degeneration (n) (1) a decrease in the extent of development of an organ, usually because of lack or loss of functional use. This can be during the life of an individual (e.g. muscles degenerate when not used) or in the evolution (↑) of a species, e.g. the tail in man's ancestors has degenerated into the coccyx. (2) death and breakdown of cells and nerve fibres.

vestigial organ (n) an organ which has become reduced in size and non-functional because it is no longer needed as a result of evolutionary changes, e.g. birds which no longer need to fly have vestigial wings.

radioactive dating (n) a method of estimating the age of prehistoric sediment or fossils from the amount of naturally occurring radioactive isotopes still present, e.g. carbon dating measures the amount of carbon -14 remaining.

geological time, geological time scale (*n*) system of sub-dividing the period of time involved in the evolution of living things on earth. The main divisions are **eras,** each of which is further divided into **periods,** e.g. Proterozoic, Palaeozoic, Mesozoic, Caenozoic (*see diagram*). The approximate dates of the eras have been obtained by radioactive dating of rocks and by identification of the fossilised remains of plants and animals that are contained within many of them.

**divisions of
geological time**

ERA	PERIOD	EPOCH	MILLIONS OF YEARS AGO
CAENOZOIC	Quaternary	Recent (Holocene)	0·01
		Pleistocene	2
	Tertiary	Pliocene	7
		Miocene	26
		Oligocene	38
		Eocene	54
		Palaeocene	65
MESOZOIC	Cretaceous		135
	Jurassic		190
	Triassic		225
PALAEOZOIC	Permian		280
	Carboniferous*		345
	Devonian		395
	Silurian		440
	Ordovician		530
	Cambrian		570
PROTEROZOIC	Precambrian		1850
ARCHAEOZOIC			3500
AZOIC			4600

*In America, the Carboniferous period is divided into the Pennsylvanian and Mississippian epochs.

climate tropical rain forest
hot and wet

high
tree
canopy

low
tree
canopy

vines and
palm trees
with orchids
and commensal
plants on
their trunks

few shrubs +
herbaceous
plants because
of tree canopy

climate

desert
vegetation
very sparce
(water is the
limiting factor)

climate
hot and dry

cactus, a
xerophyte

sand

biome

biosphere (*n*) the part of the earth and its atmosphere (i.e. land, sea and air) inhabited by living organisms.

environment (*n*) term describing the sum total of the conditions in which an organism exists, e.g. water, temperature, light, other living organisms. Growth, development and behaviour are all affected by it. This term is sometimes referred to as the external environment to distinguish it from the internal environment (p.11).

ambient (*adj*) concerned with the surroundings of an organism, i.e. its environment (↑); e.g. ambient temperature is the temperature of the surroundings.

biotic environment (*n*) the part of the environment (↑) of an organism that results from its interrelationship with other organisms; e.g when a farmer cuts down hedgerows and trees he alters the biotic environment of the animals and plants in that area.

climate (*n*) type of weather in a particular area. It is described in terms of factors such as temperature, amount of rainfall, humidity. Climatic factors are part of the environment (↑) of an organism.

biome (*n*) a large natural area that has a particular climate or physical conditions, e.g. desert, rain forest, tundra, grassland, mountain. The flora (↓) and fauna (↓) of each biome are adapted to the conditions in it.

flora (*n*) (1) the plants found in a particular area, a specific environment or a period of time; (2) a list and description of the plants found in a particular area arranged in families, genera and species.

fauna (*n*) the animals found in a particular area, specific environment or period of time.

habitat (*n*) the locality in which a plant or animal lives, e.g. pond, seashore, hedgerow. It has a particular environment (↑) to which the organism is adapted.

niche (*n*) (1) locality occupied by an organism in its habitat (↑) (2) level or position of importance of an organism within its community.

territory (*n*) a particular part of a habitat (↑) inhabited by an animal or family group of animals. Any intruders are attacked, especially just before or during the mating season. Fish, reptiles, birds and mammals often have their own territories. Territorial behaviour ensures that too many animals do not live and breed in one area.

indigenous (*adj*) describes an animal or plant that lives naturally in a particular area. It excludes those artificially introduced by man.

wild type (*n*) type of individual found most often in a wild population. This contrasts with varieties developed by man from artificial breeding programmes.

limiting factors (*n*) the particular environmental factor (p.227) restricting the growth or activity of an organism by being present at too low a level, e.g. water is the limiting factor for plants in droughts or desert habitats.

euryhaline (*adj*) describes: (1) plants which can tolerate wide variations in osmotic pressure; (2) plants or animals which can tolerate variation in the salinity of water and can therefore be found in both estuaries and in the open sea. *Contrast stenohaline* (↓).

stenohaline (*adj*) describes: (1) plants unable to tolerate variations in osmotic pressure; (2) plants or animals which can only tolerate very small variations in the salinity of water. *Contrast euryhaline* (↑).

eurythermous (*adj*) describes organisms which can tolerate variations in the temperature of their environment. *Contrast stenothermous* (↓).

stenothermous (*adj*) describes organisms unable to tolerate wide variations in the temperature of their environment. *Contrast eurythermous* (↑).

halophyte (*n*) a plant that can grow in salty soil, e.g. that found on shores, banks of estuaries and in salt marshes. It has the characteristics of a xerophyte (↓), e.g. succulence, but differs in being unable to survive prolonged drought.

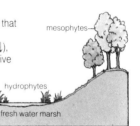
habitats of hydrophytes, mesophytes and halophytes

hydrophyte (*n*) a plant that grows in water or very wet places like freshwater ponds and marshes, e.g. water lilies. They can be free floating or rooted.

mesophyte (*n*) a plant that grows in places where the water supply is adequate, e.g. most angiosperms.

xerophyte (*n*) a plant that lives in dry habitats such as deserts and can withstand prolonged drought, e.g. cacti. Adaptations enabling it to survive in these conditions are called **xeromorphic characteristics** and may include: rolled leaves; leaves reduced to spines; leaves with thick cuticle; waxy leaves; shedding of leaves when water is lacking; sunken stomata; water storage cells.

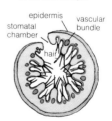
rolled leaf of **xerophyte**

ecosystem (*n*) an ecological unit formed by the interactions of the organisms within a community both with themselves and with the non-living environment, e.g. a grassland, seashore. An ecosystem has four components: (1) the producers which are autotrophs, mainly green plants; (2) the consumers which are heterotrophs, mainly animals; (3) the decomposers which are heterotrophs, mainly bacteria and fungi; (4) the non-living environment. The nature of the organisms in the ecosystem is determined mainly by environmental conditions. The activities of the organisms may, over a period of time, alter the environment.

community (*n*) in ecology, naturally occurring group of different plant and animal species that live in a particular environment. The individuals within a community interact with each other and with the non-living environment. Large communities, e.g. forests, contain many smaller communities. A community together with the non-living environment makes up an ecosystem (↑).

association (*n*) a very large plant community (↑) or communities containing normally only one or two dominant species and named after them, e.g. oak-beech association of a deciduous forest.

consociation (*n*) a plant community (↑) where one species is dominant, e.g. an oak wood. Several consociations may be linked to form a large association (↑).

population (*n*) the number of plants or animals of the same type, usually of the same species, that live in a given area.

cline (*n*) a continuous gradation of characteristics in a population of a species from one locality to another. It is often the result of variations in environmental factors such as soil, climate.

climax (*n*) in ecology, the last in the successive series of plant and animal communities (↑) which develop during colonisation of a particular area. Provided the environment does not alter this will be maintained as a steady state condition.

colony (*n*) (1) a group of animals or plants, especially of one species, living in a particular area. Individuals in some colonies, especially insects, e.g. ants, show a high degree of social organisation. In other cases they are joined structurally, e.g. sponges. (2) a culture of micro-organisms.

succession (*n*) in plants, the sequence of changes in the species of plants present from the first colonisation of an area to the establishment of a climax community (p.229). Algae and lichens are normally the first plants established, followed by mosses, then higher plants.

sere (*n*) a particular succession (↑) of plant communities that end in the establishment of a climax community (p.229). Various types of sere originate in different environments, e.g. **hydroseres** originate in water, **xeroseres** in dry conditions, **lithoseres** on rock surfaces, **haloseres** in salt marshes.

society (*n*) in ecology, a small plant climax community (p.229) found inside a larger ecological community.

dominant (*adj*) (1) in ecology, describes the major plant species in a community (p.229), e.g. oak trees in an oak wood. The dominant species determines the type and number of other species in the community. If there is more than one dominant species they are called **co-dominants**. (2) describes the top individual in a group of animals.

competition (*n*) demands made by members of a community (p.229) for essential requirements available only in limited quantities, e.g. animals compete for food.

commensalism (*n*) an association between two or more different species of organism where neither one is noticeably affected, e.g. plants that grow on tree trunks. *Contrast symbiosis* (p.30).

mutualism (*n*) an association between two organisms from which each benefit, i.e. symbiosis (p.30); e.g. certain hermit crabs have a sea anemone on the shell. The crab transports and provides food for the anemone; the anemone's sting protects the crab.

mimicry (*n*) similarity in appearance between animals of different species; it provides protection from predators; e.g. one species which might otherwise be attacked resembles another species which is not, often because the latter is poisonous; predators therefore leave both species alone. Mimicry is very common in insects.

Batesian mimicry (*n*) mimicry (↑) in which one of the species is poisonous or distasteful to predators; the harmless species is protected because of its similarity to the harmful species.

Müllerian mimicry (*n*) mimicry (↑) in which both species are harmful or distasteful to predators; since they resemble each other predators avoid both species.

succession

1–10 years · grassland

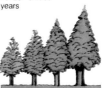

10–25 years · shrubs

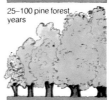

25–100 years · pine forest

100+ years · hardwood forest

succession of vegetation in an abandoned field

sea anemone

hermit crab

mutualism
mutualism between sea anemone and hermit crab

zone

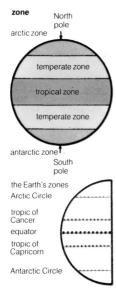

North pole

arctic zone

temperate zone

tropical zone

temperate zone

antarctic zone

South pole

the Earth's zones

Arctic Circle

tropic of Cancer

equator

tropic of Capricorn

Antarctic Circle

soil

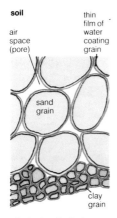

thin film of water coating grain

air space (pore)

sand grain

clay grain

soil, showing distribution of air and water

region (n) (1) an area of land with particular distinguishing characteristics, e.g. polar regions are distinguished by a cold climate. (2) part of an animal or plant that usually has a particular function, e.g. head region includes all structures on or near the head.

zone (n) (1) in geography, a sub-division of the earth's surface, often band-shaped (*see diagram*). Also used to describe a sub-division of a smaller area with particular characteristics, e.g. tree zone of a mountain. (2) part of a region (↑) in a plant or animal with a particular function, e.g. **zone of elongation** lies behind the growing tip of a root or stem.

transect (n) sample belt or a line drawn across a region (↑) of vegetation which is thereby marked off to study changes in composition of vegetation.

quadrat (n) randomly selected square (usually one metre square for ground flora) of vegetation selected to study the plants present in a region.

terrestrial (adj) concerned with the ground. Animals (e.g. cattle) and plants (e.g. maize) have terrestrial habitats.

arboreal (adj) concerned with trees, e.g. monkeys have arboreal habitats.

rainfall (n) amount of rain, in cm, that falls in a given time at a particular place. Rainfall in rain forests is high, that in deserts is low.

soil (n) surface layers of the earth's crust consisting of grains which have water around them and air spaces between. It is formed by the weathering (p.233) of rocks. Plants grow in and obtain nutrients from soil. The type, size (the **crumb structure**) and relative proportions of the grains, i.e. sand, silt, clay, humus (p.232) etc., determine the nature of the soil and hence the type of plants which grow in it. The presence of micro-organisms in soil affects **soil pH**; most plants grow best in soils of pH6-7.

soil profile (n) layers of different types of soil visible in a vertical section through the surface layers of the earth. A soil profile gives information on the character of a soil.

top soil (n) top layer of soil (↑) which is usually thinner than the underlying subsoil (↓). Plants obtain nutrients from it. If it is removed, e.g. by erosion (p.233), soil fertility is much reduced.

subsoil (n) layer of soil (↑) lying between the top soil (↑) and the underlying rock. Less fertile than top soil.

rock (*n*) hard, solid material of the outer part of the earth's crust. In many places it is covered by subsoil (p.231) and top soil (p.231) which have been formed by its weathering (↓). Various types of rock are formed in different ways and consist of one or more minerals; some contain remains of dead plants and animals, e.g. igneous rock, sedimentary rock, metamorphic rock.

chalk (*n*) white, soft rock (↑) composed of calcium carbonate; formed from shells of sea animals deposited on the sea bed.

calcareous (*adj*) chalky (↑) in nature.

limestone (*n*) medium-hard rock (↑) composed of calcium carbonate, sometimes with much magnesium carbonate. It is formed from sediment on shallow sea beds.

sand (*n*) mass of separate grains formed by weathering (↓) of rock (↑). Pure sand is mainly silica or silicates. Sandy soil is very porous (↓).

gravel (*n*) rock (↑) particles which are larger than those of sand (↑) but smaller than those of stone.

clay (*n*) mass of very fine particles formed by the action of carbonic acid on certain rocks. It consists mainly of aluminium silicates and is almost impervious (↓) to water.

silt (*n*) fine, mud-like sediment (↓). Its grains are larger than those of clay (↑) but smaller than sand (↓).

sediment (*n*) solid particles that separate from a liquid, e.g. particles of clay, sand and gravel settle, by the action of gravity, to the bottom of a stream when the current becomes too slow to keep them suspended.

humus (*n*) organic matter found in top soil (p.231); formed by decomposition of plant and animal tissues. Important for plant growth as it contains many minerals, e.g. nitrates; it also helps the soil retain water.

loam (*n*) soil consisting of a natural mixture of sand and clay with humus. Most of its physical and chemical characteristics are intermediate between those of clay and sandy soils. It is the best soil for agriculture.

fertile (*adj*) describes (1) a soil rich in minerals and on which crop yields are high; (2) plants and animals with seeds, eggs or female gametes than can develop after fertilization. *Contrast infertile* (↓), *barren* (↓). **fertility** (*n*).

infertile (*adj*) describes (1) a soil that is poor in minerals and on which crop yields are low; (2) plants and animals that cannot produce offspring. *Contrast fertile* (↑), *barren* (↓). **infertility** (*n*).

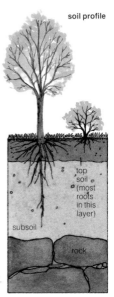

soil profile

top soil (most roots in this layer)

subsoil

rock

soil profile
Note only plants with deep root systems penetrate to the subsoil.

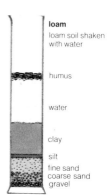

loam
loam soil shaken with water

humus

water

clay

silt
fine sand
coarse sand
gravel

mechanical separation of soil particles of different sizes

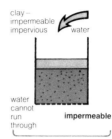

clay –
impermeable
impervious   water

water
cannot
run   **impermeable**
through

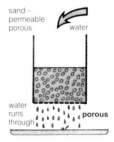

sand –
permeable
porous   water

water
runs   **porous**
through

porosity of sand and clay

strips of different
crops around hillside

contour planting

barren (*adj*) describes (1) a soil on which crops cannot be grown; (2) plants and animals that cannot produce offspring. *Contrast fertile* (↑), *infertile* (↑).

porous, permeable (*adj*) describes a material made up of loosely packed particles through which liquids can readily pass, e.g. a sandy (↑) soil allows water to drain quickly. *Contrast impermeable* (↓).

impermeable, impervious (*adj*) describes a material made up of closely packed particles that does not allow liquids to pass through, e.g. clay soil (↑) does not allow water to drain away. *Contrast porous* (↑).

edaphic factors (*n*) environmental conditions which are determined by the physical, chemical and biological nature of the soil; affect organism present.

weathering (*n*) action of wind, rain, water, ice formation, frost, chemicals and plant roots on rock. It breaks rock up into smaller and smaller particles, eventually forming soil.

leaching (*n*) the washing away, by a liquid, of soluble substances from a solid, e.g. minerals are leached from top soil as rain water passes through; soil becomes infertile (↑), few plants grow and soil erosion (↓) occurs. **leach** (*v*).

soil erosion (*n*) gradual loss of top soil from an area by physical or chemical means, e.g. by wind, surface water running down hills. This is often the direct result of destruction of the natural vegetation. **Sheet erosion** is the uniform erosion of an area's top soil; **gully erosion** is the formation of deep channels.

soil conservation (*n*) maintenance of soil fertility by prevention of erosion (↑). This is carried out by terracing (p.234), contour planting (↓), contour ridging, strip cropping (↓) and use of cover crops (p.234).

conservation (*n*) preservation of the existing state, e.g. flora and fauna are conserved so species do not become extinct; soil is conserved to retain its fertility.

contour planting (*n*) growing of crops on slopes and hillsides in strips that follow the contour lines. This prevents erosion (↑) of soil by wind and surface water running straight down the hill.

strip cropping (*n*) planting alternate bands of soil binding and non-soil binding crops in contour planting (↑). Helps prevent erosion (↑) by wind and surface water running straight down the hill.

terracing (*n*) cutting out narrow, level strips of soil around a hillside so that the soil can be cultivated.

cover crop (*n*) crop grown so that the soil surface is covered. Prevents erosion (p.233) because it is not dried by the sun and blown away by wind. Roots also bind the soil to prevent erosion by water.

crop (1) (*n*) type of vegetation being cultivated in a particular field. (2) (*n*) total amount of produce (the plant or its seeds or fruits) harvested from a particular field. (3) (*v*) to collect the produce that has been cultivated.

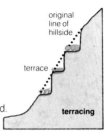

terracing

manure (*n*) (1) a substance put on soil to increase humus and mineral content. (2) animal excreta, often mixed with bedding, e.g. straw, which is used as in (1).

compost (*n*) a mixture of decayed organic substances, especially plant remains, put on soil to increase its humus and mineral content.

fertilizer (*n*) any mineral(s) added to the soil to replace those removed by crops, e.g. ammonium nitrate. Fertilizers keep the soil fertile and increase crop yields. *Contrast manure* (↑) *and compost* (↑) which replace both minerals and humus.

crop rotation (*n*) system of agriculture involving planting a field in successive years with various crops, each of which has a different nutrient requirement, e.g. legume, cereal, root vegetable, fallow year. Every 2 or 4 years the field is left fallow (↓) or used to graze cattle.

fallow (*adj*) describes land left without a crop (crop rotation ↑) so soil fertility and plant nutrients increase.

water table (*n*) level in the ground below which water cannot drain further and therefore collects.

water culture (*n*) method of growing plants with roots in a solution of plant nutrients. Often used experimentally to find the optimum mineral requirements.

hydroponics (*n*) large scale culture of plants where roots are in a solution of nutrients (water culture ↑) or are rooted in an inert material, e.g. sand supplied with water containing all the required nutrients, i.e. **sand culture**. Yields are similar to those for soil grown plants.

water cycle (*n*) the circulation of water molecules in nature. Rainfall irrigates soil and provides water for plants; excess drains into streams and rivers, some of which is used by animals as it flows to the sea. Plants, animals and exposed water surfaces give off water vapour which rises to form clouds. Clouds deposit rain over land and the cycle continues.

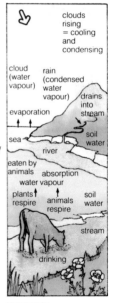

water cycle

carbon cycle

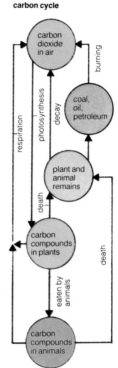

carbon cycle (*n*) circulation of carbon atoms in nature. Plants use carbon dioxide from the air during photosynthesis to build up complex organic compounds; if plants are eaten by animals the organic compounds are incorporated into animal tissues. All plants and animals release carbon dioxide into the air during respiration. When they die and decay the residual carbon compounds are oxidised by fungi and bacteria to carbon dioxide. Carbon dioxide is also released into the air by burning wood, coal, petroleum, etc.

weed (*n*) a plant growing where man does not want it to grow, e.g. amongst cultivated plants.

pest (*n*) any living organism that is a nuisance to man because it damages crops or spreads disease, e.g. locusts, fungi.

pesticide (*n*) any substance that kills pests, e.g. fungicide (↓), insecticide (↓), herbicide (↓).

fungicide (*n*) any substance that kills a fungus. **fungicidal** (*adj*).

insecticide (*n*) any substance that kills insects, e.g. **DDT. Contact insecticides** kill when they penetrate cuticle or block spiracles; **systemic insecticides** when absorbed into the body organs.

herbicide (*n*) any substance that kills plants, particularly weeds, e.g. selective weedkillers.

biological control (*n*) use of a predator to decrease the population of a pest (↑), e.g. fish are introduced into ponds to eat the larvae of mosquitos that spread disease.

blight (*n*) disease of plants, caused by virus or fungus, that makes them stop growing and wither.

mosaic (*n*) disease of plants, caused by a virus, that results in leaves having a brown, yellow or black mosaic pattern.

ergot (*n*) disease of grasses and cereals, particularly rye; caused by a fungus. Consumption of bread made from rye diseased with ergot causes poisoning.

pollution (*n*) the release of material into the environment which makes it harmful to human, animal or plant life, e.g. waste from factories pollutes the air; human faeces and heavy metals from factories pollute water.

aquatic (*adj*) concerned with water, e.g. aquatic plants and animals live on or in water.

pool (*n*) a body of still water. It is often shallow and may be present only temporarily.

eutrophic (*adj*) describes a lake or river which has a very high supply of nutrients and therefore supports abundant plant growth. This may occur naturally or be the result of pollution, e.g. by excess fertilizers that drain from the land or by industrial waste that flows into the river. *Compare oligotrophic* (↓).

oligrotrophic (*adj*) describes a lake or river that has a low supply of nutrients and therefore is not able to support abundant plant life. *Compare eutrophic* (↑).

stagnant (*adj*) in ecology, describes water in a pond, lake, stream or river that is not moving and therefore becomes foul due to decay of plants and animals.

freshwater (*n*) water found in rivers and lakes that contains only low levels of salts. *Contrast sea-water* (↓).

sea-water (*n*) water found in the sea. It contains several salts in the following approximate concentrations: sodium chloride 2.8%; magnesium chloride 0.4%; potassium chloride 0.1%; magnesium sulphate 0.2%; calcium sulphate 0.1%. *Contrast freshwater* (↑).

saline (1) (*n*) containing sodium chloride. (2) (*n*) physiological saline (p.12) which is the same concentration as blood plasma and is used in experimental investigations and to treat physiological conditions.

salinity (*n*) the amount of sodium chloride present in solution, e.g. the Dead Sea has a high salinity because it contains a high concentration of sodium chloride.

brine (*n*) (1) the sea. (2) water containing salts of similar concentrations to that of sea-water (↑).

marine (*adj*) concerned with the sea and oceans, e.g. marine organisms live in sea-water.

estuary (*n*) wide, lower part of a river where it flows into the sea. Salinity (↑) of the water varies both with state of the tide (being highest at high tide and lowest at low tide) and with distance from the sea (being highest near the sea and least at the furthest point up river to which the tide flows). **estuarine** (*adj*).

shore, beach (*n*) usually sandy or pebbly land bordering on to the sea or large expanse of freshwater.

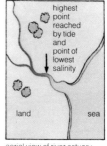

aerial view of river estuary
at high tide

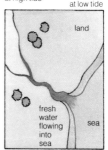

at low tide

freshwater
fresh + salt water
salt water

sea water   freshwater

tide (*n*) the twice daily ebb and flow of the sea caused by the attraction of the moon. At **high tide** water level is highest and the least amount of shore (↑) is exposed; at **low tide** water level is lowest. **tidal** (*adj*).

intertidal (*adj*) concerned with the region of the shore (↑) between the water mark at low and high tides.

benthic division (*n*) a division of an ocean or lake based on its nearness to the ocean/lake bed. Organisms present (**benthos**) are either attached to the ocean/lake bed or crawl about over it. *Contrast pelagic division* (↓).

pelagic division (*n*) a division of an ocean or lake based on its distance from the ocean/lake bed. Organisms present are free swimming, i.e. not attached to the bottom, e.g. plankton, nekton (p.36). *Contrast benthic division* (↑).

oceanic province (*n*) a division of the ocean based on depth of water. It is deeper than 200m, i.e. beyond the continental shelf (↓). *Contrast neritic province* (↓).

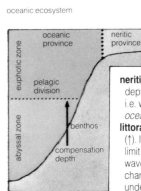

oceanic ecosystem

littoral zone

land

oceanic province

neritic province

benthos

euphotic zone

continental shelf

pelagic division

abyssal zone

benthos

compensation depth

neritic province (*n*) a division of the ocean based on depth of water. Its water is not more than 200m deep, i.e. water of a shallow continental shelf. *Contrast oceanic province* (↑).

littoral zone (*n*) (1) a sub-division of the neritic province (↑). It extends from the high tide mark outwards to the limit of the water that is completely stirred by tides and waves. Distribution of plant and animal species changes rapidly in passing from the part permanently under water to that exposed at low tide. Littoral organisms form part of the benthos (↑). (2) area of a lake from the water's edge to 6-12m, i.e. where light penetrates.

abyssal zone (*n*) division of an ocean or lake based on penetration of light. It is the deep waters where light does not penetrate and photosynthetic organisms are therefore absent. *Contrast euphotic zone* (↓).

euphotic zone, photic zone (*n*) division of an ocean or lake based on penetration of light. It is the surface waters where light penetrates and photosynthetic organisms are thus found. *Contrast abyssal zone* (↑).

compensation depth (*n*) the water between the abyssal (p.237) and euphotic (p.237) zones where the rate of respiration of plants equals their rate of photosynthesis. The exact depth depends on the turbidity of the water and also on intensity of sunlight.

free swimming (*adj*) describes a single organism that swims about freely and is not attached either to other individuals or other objects, e.g. fish.

free drifting (*adj*) describes a single, unattached organism that is non-motile and is carried about by water currents, e.g. plankton.

free living (*adj*) describes a single animal that lives separately, not attached either to other individuals of the same species or other objects, e.g. man.

atom (*n*) smallest particle of an element that shows all the properties of the element. It consists of a nucleus and electrons.

molecule (*n*) smallest group of atoms of an element or compound which has a free existence; e.g. an oxygen molecule (O_2) consists of two oxygen atoms joined together by a chemical bond (↓); a carbon dioxide molecule (CO_2) is one carbon atom joined to two oxygen atoms.

macromolecule (*n*) a very large molecule formed from a number of simpler molecules, e.g. proteins formed from amino acids, carbohydrates from sugars, nucleic acids from nucleotides.

substance (*n*) a material that has a definite chemical composition which is constant whatever its source, e.g. common salt always has the formula NaCl.

isomer (*n*) one of two forms of a chemical compound. Both have the same type and number of atoms in their molecule but have a different arrangement. Physical and chemical properties of the two forms are different.

ion (*n*) a single atom or group of atoms chemically combined that has gained or lost an electron and so has an electric charge. **Cations** are formed by the loss of an electron(s) and therefore have a positive charge, e.g. Na^+. **Anions** have gained an electron(s) and therefore have negative charge, e.g. Cl^-. Ions in compounds are held together by ionic bonds (↓).

chemical bond (*n*) a force by which the atoms or ions (↑) that make up molecules and crystals are held together. An **ionic bond** occurs between two ions, e.g. in a sodium chloride molecule the sodium atom loses an

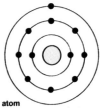

atom
sodium atom

electrons in orbits around the nucleus

molecule
carbon dioxide molecule

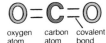

oxygen atom

carbon atom

covalent bond

formula

glyceraldehyde

isomer

CH₂OH
|
CHOH
|
CHO

CH₂OH
|
CO
|
CH₂OH

dihydroxyacetone

ion
common ions in biology

cations	anions	
H⁺		
Na⁺	OH⁻	hydroxyl
K⁺	Cl⁻	chloride
NH₄⁺	NO₃⁻	nitrate
Mg²⁺	SO₄²⁻	sulphate
Ca²⁺	CO₃²⁻	carbonate
Cu⁺, Cu²⁺	PO₄³⁻	phosphate
Zn²⁺	COO⁻	carboxyl
Fe²⁺, Fe³⁺		

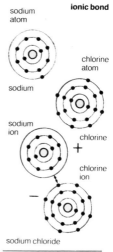

sodium atom

ionic bond

chlorine atom

sodium

sodium ion

chlorine

+

chlorine ion

−

sodium chloride

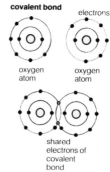

covalent bond

electrons

oxygen atom

oxygen atom

shared electrons of covalent bond

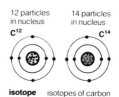

12 particles in nucleus

14 particles in nucleus

C^{12}

C^{14}

isotope isotopes of carbon

electron and becomes a positively charged sodium ion (Na^+); the chlorine atom gains an electron and becomes a negatively charged chlorine ion (Cl^-); attraction between the oppositely charged ions is the ionic bond. A **covalent bond** is between two atoms and involves sharing a pair of electrons, e.g. an oxygen molecule is made up of two oxygen atoms joined by a covalent bond.

hydrogen bond (*n*) a weak type of ionic bond (↑) of particular importance in living tissues, e.g. hydrogen bonds between the base pairs hold together the DNA molecule.

hydrophobic (*adj*) describes a chemical that repels water and is insoluble in it, e.g. hydrocarbons including fats and oils. *Contrast hydrophilic* (↓).

hydrophilic (*adj*) describes a chemical that attracts water and is soluble in it, e.g. substances that split into ions like potassium chloride. *Contrast hydrophobic* (↑).

radioactivity (*n*) spontaneous disintegration of atomic nuclei; first observed in heavy elements like radium and uranium. Radioactive elements or compounds give off certain rays ($\propto$, ß or γ) when this occurs.

irradiation (*n*) exposure to light rays or other types of ray like the $\propto$, ß or γ rays from radioactive (↑) material.

radioactive tracers (*n*) radioactive isotopes (↓), e.g. C^{14}, used for experimental or diagnostic purposes to follow particular atoms or compounds in a complex mixture of reactions; e.g. small doses of radioactive amino acids can be administered to animals to study protein metabolism. Estimated by using a Geiger-Müller tube or scintillation counter.

isotope (*n*) certain atoms of an element which differ from ordinary atoms only in the mass of their nuclei. Detected by scintillation counter if radioactive, e.g. C^{14} (carbon[14]), or by mass spectroscope if non-radioactive, e.g. C^{12} (carbon[12]).

type of wave radio	infra-red	visible	ultra-violet	x-ray	gamma-ray
1000m →1cm	10^{-4}cm	10^{-5}cm	10^{-6}cm	10^{-8}cm	10^{-9}cm

electromagnetic waves

radiation sickness (*n*) illness due to excess exposure of the body to radioactive material (p.239).

X-rays (n) types of electromagnetic wave with a very short wavelength. They can penetrate matter which is opaque to light; e.g. they pass differentially through tissues and can be used to obtain X-ray pictures of internal body organs (e.g. bones) and are therefore useful in medical diagnosis.

infra-red rays, i.r. light (*n*) types of electromagnetic wave with wavelengths longer than those of the red light in the visible spectrum.

ultra-violet rays, u.v. light (*n*) types of electromagnetic wave with wavelengths shorter than those of violet light in the visible spectrum. The browning effect on the skin in strong sunlight is due to them, *see melanin* (p.157).

optimal conditions (*n*) the best conditions possible, e.g. **optimal temperature** for enzyme activity (37°C) is that at which enzyme reactions proceed most rapidly.

pH value (*n*) value on a numerical scale, between 0 and 14, which expresses the acidity or alkalinity of a solution. At pH 7 the solution is neutral, *see diagram*. pH value can be determined by use of indicators (↓).

acid (*n*) chemical compound which when dissolved in water gives a solution with a pH value (↑) below 7. Acids turn litmus (↓) red. **acidic** (*adj*).

alkali (*n*) chemical compound which when dissolved in water gives a solution with a pH value (↑) above 7. Alkalis turn litmus (↓) blue. **alkaline** (*adj*).

indicator (*n*) a substance that is a different colour in acidic and alkaline solutions; used to determine pH value (↑). Common indicators include **universal indicator, litmus, phenolphthalein, methyl orange**.

buffer (*n*) in science, a solution resistant to a major change in pH (↑) on addition of acid or alkali. A mixture of a weak acid and one of its salts, e.g. ethanoic (acetic) acid and its sodium salt, will act as a buffer. Carbonic and phospheric buffers are important in controlling the pH of fluids in living organisms.

reagent (*n*) a substance which when added to another substance(s) causes a chemical reaction.

salt (*n*) chemical compound formed when the hydrogen in an acid is replaced by a metal; e.g. the salt calcium sulphate ($CaSO_4$) is formed when the metal calcium (Ca) replaces the two hydrogen atoms in a molecule of sulphuric acid (H_2SO_4).

pH scale

pH value

indicator

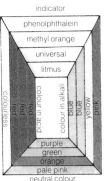

neutral colour

benzene ring

abbreviated form
for benzene ring

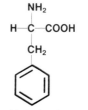

NH₂

H—C—COOH

CH₂

phenylalanine formula
aromatic compound

CH₂OH   alcohol

CH₃   ethyl alcohol
            (ethanol)

carboxylic acid
carboxyl group   —C
            O
            OH

CH₃—C
            O
acetic acid
            OH

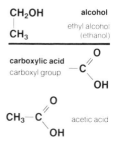

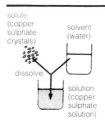

solute
(copper
sulphate
crystals)   solvent
(water)

dissolve

solution
(copper
sulphate
solution)

solution
formation of a solution

acid salt (*n*) chemical compound formed when part of the hydrogen in an acid is replaced by a metal; e.g. the acid salt sodium acid carbonate or sodium bicarbonate (NaHCO₃) is formed when sodium (Na) replaces one hydrogen atom in carbonic acid (H₂CO₃).

benzene ring (*n*) structure of a benzene molecule. It is found in many other compounds, e.g. phenylalanine.

chelate (*n*) complex compound in which a metal ion is attached to an organic molecule, e.g. haemoglobin.

aromatic compound (*n*) organic compound containing a benzene ring (↑), e.g. tyrosine, phenylalanine.

aliphatic compound (*n*) organic compound which does not contain a benzene ring (↑), e.g. a fat.

alcohol (*n*) organic compound that contains one or more hydroxyl (−OH) groups, e.g. ethyl alcohol.

ester (*n*) organic compound formed from an alcohol (↑) and an acid (usually an organic acid). This type of reaction is called **esterification.**

carboxylic acid (*n*) an acid which contains a carboxyl group (−COOH), e.g. acetic acid.

polymer (*n*) macromolecule (p.238) made up of a large number of similar small molecules (called **monomers**) joined in a similar way, e.g. nucleotides join to form nucleic acids. Reactions are called **polymerisations**.

straight chain molecule (*n*) organic molecule in which the chain of carbon atoms is straight; no carbon atom is attached to more than two other carbon atoms. *Contrast branched chain molecule* (↓).

branched chain molecule (*n*) organic molecule in which the chain of carbon atoms is branched; some of the carbon atoms may be attached to three or four other carbon atoms. *Contrast straight chain molecule* (↑).

solution (*n*) result of dissolving (↓) a solid or gas (the **solute**) in a liquid (the **solvent**). The solvent is usually specified, e.g. fat is soluble in ethanol; if not specified it is taken to be water. Solutions are concentrated (↓) or dilute (p.242), saturated or unsaturated (p.125). **Solubility** is the maximum amount of solute which can be dissolved at a given temperature.

concentration (*n*) amount of solute dissolved in solution; measured in grammes/vol or moles/vol, e.g. 10g/dm³.

dissolve (*v*) to disperse a substance into a liquid (solvent) to form a solution (↑).

concentrated (*adj*) describes a solution (↑) that contains a lot of solute. *Contrast dilute (p.242).*

dilute (*adj*) describes a solution (p.241) that contains a little solute. *Contrast concentrated* (p.241).

soluble (*adj*) describes a substance that will dissolve in a named solvent, e.g. glucose is soluble in water. *Contrast insoluble* (↓).

insoluble (*adj*) describes a substance which will not dissolve in a named solvent, e.g. fats are insoluble in water. *Contrast soluble* (↑).

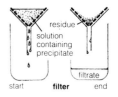

filter (*n*) a device for separating a solid from a liquid. It consists of a porous material through which the liquid but not the solid can pass, e.g. a precipitate (↓) is filtered using **filter paper** (*see diagram*). The liquid which passes through is the **filtrate,** the solid is the **residue,** the process is filtration (p.153). **filter** (*v*).

precipitate (*v*) (1) in chemistry, to come out of solution; e.g. dilute sulphuric acid added to lead nitrate forms insoluble lead sulphate which is precipitated. The process is **precipitation,** the insoluble product is the **precipitate.** (2) in ecology, to fall as rain.

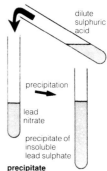

precipitate

diffusion (*n*) process by which molecules in contact with each other as gases or in solution spread into one another and form a homogeneous mixture; e.g. carbon dioxide in expired air rapidly diffuses into the atmosphere. **diffuse** (*v*).

dialysis (*n*) process by which small molecules are removed from a solution containing a mixture of small and large molecules. It involves use of a dialysis membrane which has pores of a size such that only the small molecules can diffuse through it. **dialyse** (*v*).

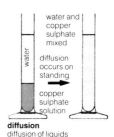

diffusion
diffusion of liquids

hydrolysis (*n*) chemical breakdown of a substance by water. Reaction is slow; acids, alkalis and enzymes speed it up. **hydrolyse** (*v*).

dehydration (*n*) removal or loss of water, e.g. blue copper II sulphate when heated loses water of crystallisation; plants dehydrate during a drought. **dehydrate** (*v*).

reversible reaction (*n*) chemical reaction that can be made to go in either direction by altering the conditions, e.g. haemoglobin + oxygen ⇌ oxyhaemoglobin. The direction depends on the concentration of oxygen.

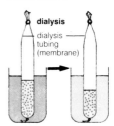

irreversible reaction (*n*) chemical reaction that will only go in one direction.

decomposition (*n*) (1) in chemistry, breakdown of a chemical compound into simpler compounds or into its constituent elements; e.g. sodium nitrate when heated

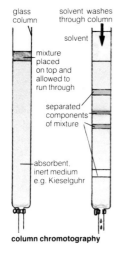

column chromotography

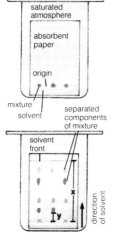

paper chromatography

decomposes to sodium nitrite and oxygen. (2) in biology, decay (↓) of organic matter; e.g. animal and plant remains decompose in soil. **decompose** (v).

decay (n) (1) rotting of an animal and plant remains, brought about by bacteria and fungi in soil leading to liberation of carbon dioxide, ammonia and water. (2) disintegration of radioactive substances.

electrophoresis (n) in biology, method of separating charged molecules by subjecting them to an electric field. The mixture to be separated (e.g. proteins) is applied to a porous medium like starch, silica gel, damp filter paper. Positive and negative ions move to the cathode or anode respectively; the rate of movement varies with size, shape and number of positive or negative charges.

chromatography (n) methods for separating a mixture of substances in solution. See column (↓), paper (↓) and thin-layer (↓) chromatography, gel filtration (↓). All depend on differential rates of movement or release after absorption as solvent moves through mixture.

Rf values (n) in chromatography (↑), rate or amount of movement of substance over amount of movement of solvent, i.e. $\frac{y}{x}$ (see diagram). All values 1.0 or less.

column chromatography (n) chromatography (↑) which uses a column of suitable material, e.g. Kieselguhr. The mixture to be separated is poured in at the top and is absorbed as it passes through. The column is then washed by a solvent which differentially extracts substances in the mixture bringing about separation.

gel filtration (n) column chromatography (↑) which uses a column of insoluble gel, e.g. Sephadex. Large molecules pass through more rapidly than smaller molecules thus bringing about a separation.

paper chromatography (n) type of chromatography (↑). A spot of the mixture to be separated is placed at the edge of a strip or sheet of paper similar to filter paper. When dry, this edge is dipped into a solvent; movement of solvent up the paper separates the components of the mixture. Used to analyse small samples of material, particularly amino acids. Rapidly being replaced by thin-layer chromatography (↓).

thin-layer chromatography (n) type of chromatography (↑) similar to paper chromatography (↑) but much more rapid. Separation is brought about on a thin layer of material, e.g. silica gel coated on a glass or plastic plate.

centrifugation (*n*) method of separating suspended particles or substances of different density by subjecting them to high forces of gravity by spinning them rapidly in a machine called a **centrifuge**. Solid particles or the densest layers are carried to the bottom of the tube.

ultracentrifuge (*n*) centrifuge (↑) which is capable of very fast rotation, e.g. 65,000 r.p.m., and therefore very high forces of gravity.

apparatus (*n*) any objects, instruments or machines used to carry out a scientific experiment (↓), e.g. **test tube, beaker, funnel, flask,** thermometer (↓), centrifuge (↑).

burette (*n*) graduated glass tube for measuring variable quantities of liquid that are run from it.

pipette (*n*) special type of glass tube for measuring and transferring to another container a volume of liquid. **Graduated pipettes** transfer variable volumes, **bulb pipettes** fixed volumes.

micro-pipette (*n*) pipette (↑) for measuring small volumes between 1 and 100 μl.

capillary tube (*n*) glass tube of very small internal diameter. If such a tube is placed in a liquid like water the liquid rises up. The height to which it rises is the capillary rise (as shown by h in the diagram on p.245). This capillary action (**capillarity**) is partially responsible for the upward passage of water and dissolved nutrients in the conducting vessels (i.e. xylem) of plants.

petri-dish (*n*) shallow glass dish with an overlapping cover. When sterilised and filled with an agar based medium it can be used to culture bacteria.

thermometer (*n*) instrument for measuring temperature. Most common ones in scientific laboratories are mercury thermometers. **Clinical thermometers** are used in medicine.

kata-thermometer (*n*) alcohol thermometer used to measure the cooling power of the air, e.g. to find the efficiency of a ventilation system.

wet and dry bulb thermometer, hygrometer (*n*) thermometer (↑) used to measure the humidity of the air.

autoclave (*n*) apparatus consisting of a strong-walled metal container in which conditions of high temperature and pressure can be reached. Used, e.g. to sterilise instruments and bacterial culture media.

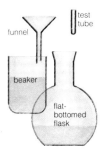

test tube

funnel

beaker

flat-bottomed flask

measuring cylinder

conical flask

apparatus

petri-dish

common, glass, laboratory apparatus

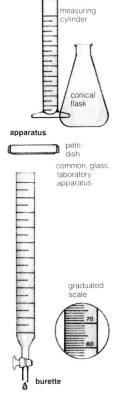

graduated scale

70

60

burette

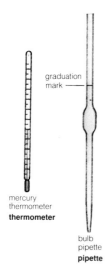

graduation mark

mercury thermometer
thermometer

bulb
pipette
pipette

capillary tube

h = capillary rise
The narrower the capillary tube, the greater the capillary rise

experiment (*n*) anything done to discover new facts or to test a theory; e.g. in scientific laboratories experiments are carried out to discover new drugs to treat disease.

in vivo (*adj*) describes experiments or biological processes that are carried out inside living organisms, especially animals; e.g. testing drugs in an infected animal is an *in vivo* experiment. *Contrast in vitro* (↓).

in vitro (*adj*) describes biological experiments carried out in laboratory apparatus; e.g. testing drugs against cultures of bacteria in test tubes is an *in vitro* experiment. *Contrast in vivo* (↑).

control experiment (*n*) a part of an experiment which tests that the effects being measured are due to the factors under consideration; e.g. when testing a solution for glucose (see Fehling's test p.115) a control experiment with water alone should be carried out to check that no colour is obtained in the absence of glucose.

observation (*n*) anything noted during an experiment, e.g. a reading on an instrument; a colour change.

identification (*n*) naming of an object or process on the basis of its properties; e.g. identification of bacteria is carried out by observing them under a microscope and subjecting them to tests (e.g. Gram's test p.27).

dissection (*n*) cutting open a plant or animal to show and study the structure of the internal organs.

test (*n*) experiment(s) carried out to: (1) see if apparatus or machinery is working correctly; (2) check certain facts are correct; (3) detect chemical substances; e.g. Biuret test to detect the presence of proteins.

analysis (*n*) in chemistry, the identification of the components of a substance or mixture. **Qualitative analysis** determines chemical composition; **quantitative analysis** determines mass. **volumetric analysis** determines the concentration of a solution.

assay (*v*) to determine the proportions of the constituents of a substance.

bio-assay (*n*) determination of the amount of a substance using biological material to measure it; e.g. antibiotics are assayed by measuring inhibition of bacterial growth produced.

sample (*n*) small part of a liquid or solid that has the same characteristics as the whole.

SI units (*n*) modern scientific system of units based on the metre-kilogram-second system.

Index